RESURRECTING
— THE —
SHARK

RESURRECTING

— THE —

SHARK

A SCIENTIFIC OBSESSION AND THE MAVERICKS WHO SOLVED
THE MYSTERY OF A 270-MILLION-YEAR-OLD FOSSIL

SUSAN EWING

PEGASUS BOOKS
NEW YORK LONDON

This book features four augmented reality models, viewable with the "Resurrecting the Shark" app, available on Google Play and iTunes. To use the app, open it on your phone or tablet and wait for the camera to activate (menu is not active until the camera is on). Point the camera at one of the target art pieces at the end of the color insert in the middle of the book, and the model will pop up on your device. As long as you keep the art in the camera frame, you can move closer to zoom in, or rotate the camera around the art to see the model from different angles. You can also view the models without the book, through the app menu. *(Model and app development led by Jesse Pruitt of the Idaho Virtualization Lab and Informatics Research Institute.)*

———————

RESURRECTING THE SHARK

Pegasus Books Ltd
148 West 37th Street, 13th Floor
New York, NY 10018

First Pegasus Books hardcover edition April 2017

Interior design by Sabrina Plomitallo-González

ISBN: 978-1-68177-343-8

10 9 8 7 6 5 4 3 2 1

Printed in the United States of America
Distributed by W. W. Norton & Company, Inc.

TABLE OF CONTENTS

INTRODUCTION

WHEN I DROVE DOWN TO POCATELLO IN JUNE 2013 TO ATTEND THE OPENING OF "THE WHORL Tooth Sharks of Idaho" exhibit at the Idaho Museum of Natural History, I expected to see illustrations and fossils of the bizarre extinct shark *Helicoprion*. I never suspected the beast would seize my imagination and shake until my brain rattled.

Truth be told, I didn't go to Pocatello for the shark. I went because I hadn't seen my artist friend Ray Troll in a long time and he was throwing one of his famous after-parties to celebrate the exhibit, which featured his artwork. There in the exhibit hall, surrounded by astonishing fossils, life-size *Helicoprion* models, and Troll's beguiling art, I started asking questions. As Ray had obsessed for more than twenty years about what the shark looked like, I began to obsess on the story— all those chance discoveries, unexpected connections, absorbing sidelights, and compelling science factoids. Walking through the exhibit I met Jesse Pruitt, a tattooed combat veteran and undergraduate student, and his adviser-turned-colleague, then–research curator Leif Tapanila. I was taken by their backstories, as well as those of the other "Team Helico" members who had just published their breakthrough findings on *Helicoprion*. The more I looked, the more I began to see the

Chinese knot of lives (animal and human, past and present), feel the undertow of deep time, and more fully appreciate the way the earth is in a constant state of change, with species rising, thriving, and disappearing.

I also happened to have a personal history with ratfish, the sole surviving representatives of the once very large and extremely diverse group of chondrichthyans (sharks and their kin) that included the dead-end branch that held *Helicoprion*. In the early 1980s, I co-owned and operated a small wooden fishing boat named *The Salty*, and my fishing partner and I hand trolled for salmon and longlined for halibut and black cod among the whales and eagles and rocky islands of Southeast Alaska. I vividly remember the first time a two-foot-long ratfish loomed up from the murky depths as I hauled in the longline. Its head broke the surface like some spectral alien being with big glassy eyes, a mutant-rabbit face, a nasty fin spine, and thin, spotted skin. I was enthralled. That ratfish laid the universe of underwater weirdness at my rubber-booted feet right there on *The Salty*'s rolling deck, and I've had a soft spot for strange sea creatures ever since. Now here was *Helicoprion*, bringing a fresh world of oceanic oddity to my attention in the landlocked museum.

We humans trace our evolutionary roots to fish. Specifically lobe-finned fish, a subset of bony fish. Sharks? No direct relation, other than the big-tent vertebrate connection. Chondrichthyans, set apart by their cartilaginous skeletons, are the bona fide Other. Not just evolutionarily. Sharks show no emotion we can read, no facial expressions we can interpret. They don't appear to play. They possess excellent hearing and sensory detection, yet have no organs for, or apparent interest in, making sounds. Much about shark behavior, physiology, and evolution remains unknown. Sharks are the strangers we can't pin down, and the puzzle we won't let go. Prehistoric sharks? Even more tantalizing. Mulling it over on the drive home to Montana, the *Helicoprion* story felt like a tale worth tracking down and telling. I quickly learned that while biological riddles supplied plenty of intrigue for the natural history account, the larger whorl-toothed shark saga was further propelled by the ever-shifting nature of science and scientific inquiry. When the tale opened in the 1880s, paleontology had just gained its foothold as a dedicated science. Today, the study of ancient life has morphed into a dynamic frontier of accessible data, ever-more-powerful technological tools, highly informed enthusiasts, chatty paleo blogs, and the blurring of once-rigid lines between science and artistic expression. Credentialed scientists rightly drive the work, but contributions come from many fronts, in many forms, with curiosity and resolve as the only nonnegotiable prerequisites.

I'm not a scientist, but it seemed like being an outsider in a story full of outsiders was as good a qualification as any. So, tying myself into that crazy knot of chance and circumstance, I began to chronicle the nineteenth-century discovery, twentieth-century confusion—and creative twenty-first-century paleontology that brought *Helicoprion* to life as never before, brilliantly executed by a quirky band of boundary-bending collaborators that looked more like a roots-rock band than ivory tower intelligentsia. At its heart, this is a story of questions and discovery, science and art, and the triumph of curiosity, passion, resourcefulness, and moxie over all else. But of course the soul of the tale is *Helicoprion* itself, the top marine predator of its time, who lived two hundred million years before *T. rex* stomped onto the scene. If dinosaurs have ever captured your imagination, prepare to fall prey to the buzz saw shark. . . .

I

LASTING IMPRESSIONS

Whilst this planet has gone cycling on according to the fixed law of gravity, from so simple a beginning endless forms most beautiful and most wonderful have been, and are being, evolved.
—Charles Darwin, *On the Origin of Species*, 1859

SCIENTISTS ESTIMATE THAT 99.9 PERCENT OF ALL SPECIES EVER TO SLINK, SWIM, FLY, fight, wail, or warble on this earth are now extinct. It's hard enough to pin down the number of species living today (one recent study pegged the wildly varying number at 8.7 million), but the number of extinct species is thought to be somewhere between five and fifty *billion*. Billion. Species. Extinct. Some of those billions made their exit in one of the five calamitous mass extinctions our planet has experienced to date. Others quietly faded away, outcompeted or unable to adapt to earth's shifting terms and conditions, long before humans arrived to give them an extra push toward the cliff. Of those estimated extinguished billions, a random few percent are chronicled in the fossil record. This is the story of one of those few—the one-in-a-billion buzz saw shark, *Helicoprion*, a species that survived over a span of some ten million years, between about 270 and 280 million years ago, long before the dinosaurs.

The world's first *Helicoprion* fossil was a curved fragment of fourteen teeth found in Western Australia in the latter part of the nineteenth century. We only know the bare-bones outline of that discovery, as none of the fleshy particulars were preserved, in either the fossil or human narrative. And there you have the maddening norm in paleontology. So much is missing. Evidence may be the scientific engine of paleontology, but we would be hard-pressed to get out of the driveway without artful reconstruction, whether the subject at hand is a scavenger-strewn fossil fish, an ancient ecosystem, an operatic museum diorama, or a set of historic circumstances.

We know the last name of the first person to stumble upon that prefatory *Helicoprion* fossil: Davis. It was definitely Mr. Davis. But his first name is lost to history, and we are left to imagine the details leading to that happenstance moment when the first human hand touched the first whorl-toothed shark, raised up from a lost world. It might well have gone like this. . . .

———

An old prospector too bent for the outback himself told Mr. Davis to look for Swan River blackbutt trees. Old-timers believed the trees, a species of tall eucalyptus with creamy white flowers, signaled the presence of gold. In his weeks of looking, Mr. Davis had found an assortment of tough-barked gums, a bit of beryl, and some quartz. That was it. Still, somehow, he felt lucky as he kicked out his breakfast fire, saddled his horse, and loaded his pack mule as dawn rose over an austere landscape the color of terra-cotta. He could get in a few good hours of scouting before the heat drove shimmering waves into the bone-dry air.

On this particular day in the early 1880s, Mr. Davis was poking up the valley of the Arthur River, a tributary of the Gascoyne River. The Gascoyne was what locals called an "upside-down" river, in that it only ran aboveground for about four months of the year and maintained a sort of hibernation flow below the dried-up riverbed the rest of the time. As Mr. Davis was about to discover, water and gold were not the only treasures the taciturn country held dear in its rocks.

The morning had been disappointing. Mr. Davis was riding slowly over the hard ground, his mind wandering perhaps to a good meal back at the pub, and perhaps some female companionship, when something caught his eye and knocked for attention: an anomalous shape, an orderly pattern, a play of light across an incised form. Squinting, he dismounted and bent to examine a split nodule of sedimentary rock that looked to him like clay ironstone. Furrowed into the exposed surface of the nodule was a gracefully curved fossil impression—fourteen serrated points fanned into an arc about the size of a woman's hand.

As a gold prospector, and therefore at the very least an amateur geologist, Mr. Davis would have known he was looking at a fossil. In the crowning years of the late nineteenth century, fossils had established themselves in the public imagination in a big way. The first life-size models of extinct animals, including dinosaurs, had been installed in 1854 with great fanfare in an exhibit at London's Crystal Palace Park. English sculptor Benjamin Waterhouse Hawkins built the models under the scientific direction of Sir Richard Owen. It was Owen who had coined the term *dinosauria*, terrible lizard, a dozen years earlier. He was a controversial figure who resembled Dickens's Scrooge in appearance and temperament, but Owen was among the leading paleontologists of the day, considered to be a brilliant naturalist and anatomist. Before the Crystal Palace Dinosaur Court exhibit opened, a posh New Year's Eve dinner party (gentlemen only) was held inside the mold for the *Iguanodon*. The eight-course meal included mock turtle soup, raised pigeon pie, and French plums. A story in the *Illustrated London News*, complete with an engraved picture of the event, reported the party as boisterous. The masses responded by flocking to the exhibit and snapping up the world's first dinosaur souvenir sets.

Steamrolling scientific progress over the next four decades would reveal how inaccurate Owen's models were and spur American paleontologist Othniel Charles Marsh to angrily deride the Crystal Palace models and scorn anyone associated with the exhibit. Of course by then, the 1890s, Marsh had plenty to be angry about. He had been ruined financially and socially, along with his bitter rival paleontologist Edward Drinker Cope, by their very public and wantonly antagonistic "Bone Wars," waged from about 1877 to 1892. As destructive, underhanded, and just plain nasty as the competitive fossil hunting and scientific feuding was, the so-called Great Dinosaur Rush that Marsh and Cope unleashed was a shot of steroids to the young science of paleontology. Before the Bone Wars, the world had a meager nine named species of dinosaurs. After the American West had been scoured by hired fossil mercenaries and the dust had settled, there were thirty-two species. Or rather, thirty-two species were eventually proven valid from the 140 species Marsh and Cope originally claimed between them from their embarrassment of specimen riches. They (mostly Marsh) named some of our most familiar and beloved dinos, like *Triceratops*, *Stegosaurus*, and *Allosaurus*—as well as *Brontosaurus*, which was one of the species that washed out. In 1903, paleontologist Elmer Riggs determined that Marsh's *Brontosaurus* was really Marsh's *Apatosaurus*, although the highly appealing Thunder Lizard persisted in dinosaur iconography for decades. But wait! In 2015, scientists released a three-hundred-page study suggesting that there really *is* a *Brontosaurus*, distinct from the *Apatosaurus*. That's paleontology for you. Don't write anything down in ink, or at least keep a store of asterisks handy. Remember that, especially in the context of ancient sharks.

But back to Mr. Davis, who probably knew about fossils and could have even seen some in his own enterprises. He might have deciphered the distinctive serrated points in the rock as shark teeth, despite the fact that the nearest tidewater was 130 miles west from where he crouched. The Gascoyne River, the longest river in Western Australia, emptied into the aptly named Shark Bay, with its mother lode of sharks and rays, along with its whales, dolphins, dugongs, turtles, and odd underwater stumps (which in 1956 would astonishingly be identified as living stromatolites, 3.5-billion-year-old life-forms to which we largely owe our oxygen atmosphere). Whatever Mr. Davis knew, or thought he knew, about the sun-warmed rock he held in his hands, he almost certainly didn't know that his fossil teeth came from a time long before dinosaurs, and were some 120 million years older than Marsh's *Apatosaurus*.

Mr. Davis's fossil-bearing nodule had heft, and the mule already carried a load, but the specimen was entrancing and possibly worth something. Maybe the mule reached his head past Mr. Davis's ear to snuffle the rock, sealing the decision to take it.

He didn't know it then, and perhaps he never did, but Mr. Davis had struck paleontological gold. Tucked in with the rest of his swag as he made his way back toward the coast was the first *Helicoprion* fossil ever collected—the initial unearthing of a prehistoric whorl-toothed shark that would vex paleontologists for more than a hundred years.

———————

In the winter of 1993, residents of Los Angeles County were talking about building arks. The year had opened with a deluge that dumped nearly twelve inches of rain in less than a week. But the heavens were at rest and it was a pleasant if cloudy January day when Alaskan artist Ray Troll and his writer friend Brad Matsen pushed through the heavy glass doors of the Natural History Museum of Los Angeles County. The building's façade still bore gunshot holes from the previous spring's riots that followed the Rodney King verdict, but down in the subterranean collections, the museum was an ocean of calm under flickering fluorescent lights.

Troll and Matsen were collaborating on a book about ancient oceans and had come to see fossil fish expert J. D. Stewart. Stewart hailed from Kansas, and Troll had gone to junior high, high school, and college in Wichita, which was connection enough for Troll to cold-call him and wangle an invitation to come nose around the museum's collections. This proclivity for picking up the phone and calling scientists out of the blue was about to send Troll down a path deep into an *atlas obscura* of real sea monsters.

Troll's current path, walking down the stairs to the collections with Matsen and Stewart, had already meandered through its share of foggy valleys, surprise turns, and sidetracks. In 1977, after getting his bachelor of fine arts in printmaking from Bethany College in Lindsborg, Kansas, Troll beat it out of the heartland to Seattle, where he grazed his way through a few government-funded art jobs, waited tables, worked at an IRS call center, and put his college degree to work as a silk screen tech, pulling thousands of T-shirts with messages like KISW ROCK! After a few years of that, he started graduate school at Washington State University, set in the Palouse wheat country with its undulating dunes of ancient soil and amber waves of grain.

In 1981, Troll dragged his newly minted master of fine arts degree to a gig teaching art workshops at a remote Coast Guard LORAN radar station on the Bering Sea north of Nome, Alaska. His brother and two sisters were living in Alaska by then, so it was a great way to see the place for himself. His sister didn't have to ask him twice to come to Ketchikan, in Southeast Alaska, and work in her seafood shop.

In the 1980s, Ketchikan was a small logging and fishing town with fewer than nine thousand residents and an average rainfall of 153 inches per year. Moss grew on the rooftops, and down at the commercial harbor, working boats creaked and jingled in the wakes of other working boats coming and going, in a lifeblood scent of diesel fuel, fish, and salt water. When Troll showed up, his sister put him to work peddling the catch of the day from a kiosk on the docks, which allowed him to exercise his natural talent for schmoozing and fed his affection for fish. Fish first caught Troll's attention as a child watching bluegills circle in a bucket in upstate New York, where he was born. Fish were his connecting narrative thread through a series of Air Force postings, until his father left the service and resettled the family in Kansas. In a transcendental moment as a moody, preteen military dependent in Puerto Rico, young Troll was sitting alone on a rock, staring out at the Caribbean, when a giant stingray surfaced and skimmed past. He took it as a sign, a sure messenger from the underwater world that he belonged with them, the fish tribe.

As the mustachioed hippie with the soulful eyes slung seafood on the Ketchikan docks, he befriended chapped-face commercial fishermen and rubber-booted fisheries biologists who told him stories, took him out on their boats, and showed him fish. Whenever he came upon a new species, he wanted to know more about it, like the first really big fish he caught, a giant red snapper. Only it wasn't a snapper, he learned, but a yelloweye rockfish (*Sebastes ruberrimus*), which he further learned has one of the longest life spans of any fish on the planet. His dead *ruberrimus* could have been a hundred years old, and was probably a female. The experience of catching that big, beautiful fish, then discovering the details of its life, sent him over the waterfall in a barrel of pride and shame. His artistic subject matter shifted from nudes, apocalyptic cityscapes, frogs in jars, and a few scattered fish, to fish. Fish, fish, fish.

Brushing off his printmaking skills, he and a friend silk-screened two hundred T-shirts with the cheeky message LET'S SPAWN to hawk at a summer seafood festival, and sold out in two days. That fall he met another artist, Michelle, who gave him a new focus, along with, in due time, a daughter and a son. Artist dads have to make a living, and since the T-shirts had proven to be such a hit, he designed some more, and he and Michelle, as full partners, dove into retailing, then wholesaling. In 1987, he created the image that would raise him to cult status in Alaska and beyond: a human skull over a crossbones of salmon, with little naked men and women forming the side borders, and the slogan SPAWN TIL YOU DIE. In 1992, the Trolls opened the Soho Coho Gallery ("Better Living Through Difficult Art") in a historic brothel perched over a salmon-spawning stream on Ketchikan's Creek Street. He released a riptide of new T-shirt and poster designs, most with a punny fish theme.

Salted here and there among the fish were dinosaur images. Before Alaska, before fish—before even first grade—there were dinosaurs, at least in Troll's personal stratigraphy. The first drawing he remembers was of a dinosaur, when he was four years old and living on an Air Force base in Japan. "Dinosaur" was one of the first words he learned to spell, latching on from there to multisyllabic magic words like "Triceratops" and "Stegosaurus." From Japan, the family transferred to Olmsted Air Force Base in Middletown, Pennsylvania. In his elementary classroom at the Seven Sorrows of the Blessed Virgin Mary, little Ray raised his hand when Sister Mary James was teaching about Noah and the Flood. "What about the dinosaurs?" he asked. The sister explained that dinosaurs were not part of God's plan. Oh. "Well, how about the plesiosaurs?" Other boys played baseball, Ray and his only brother, Tim, played museum, encouraged by their history-buff grandfather.

Now here he was, almost forty years later, playing museum with Brad and J. D. at the Natural History Museum of Los Angeles County. The NHM vertebrate paleontology department has more than 150,000 cataloged vertebrate fossil specimens spanning 450 million years of evolution. Which is a rockin' lot of evolution. In that 450 million years, animals with backbones branched out and diversified with such innovative and consequential developments as jaws, teeth, internal skeletons, limbs, lungs, hips, wings, and, for better and worse, big brains and opposable thumbs. Near the end of the day, after hours of scavenging through drawers and shelves and crates, Stewart was showing Troll and Matsen a *Dolichorhynchops*, a Cretaceous plesiosaur that lived about seventy-five million years ago. In an afterthought that would wobble the orbit of Troll's world, Stewart hunkered down and wrestled one last fossil-bearing boulder from the shadows of the bottom shelf. "Check this out," said one Kansas boy to the other.

At first glance it looked like an ammonite, an extinct subclass of spiral-shelled cephalopod—*cephalo* meaning "head," and *pod* meaning "feet," a nod to the way their arms are attached directly to their heads. Cephalopods have eight or more arms, and sometimes an additional two or more tentacles for grasping prey. The living classmates of this long-surviving group of intriguing invertebrates includes squid, octopi, and nautiloids. What was it doing in with the vertebrates?

Then Stewart said, "Look closer. Those are *teeth*, man! It's a coil of teeth, from a bizarre extinct shark that has blown paleontologists' minds for a century. You never see it in exhibits because nobody knows how to reconstruct the freaking thing."

Troll was smitten. Spellbound. Embedded in the chocolate-brown chunk of stone was not only a very unusual fossil but also a set of elements irresistible to a curious fine artist increasingly infatuated with the fantastical side of science. First, there was the spiral, that mystical, foundational pattern repeated across nature, culture, and the cosmos from snails to petroglyphs to galaxies. (This particular fossil had a full spiral, not just a partial arc of teeth like Mr. Davis's specimen.) And, yikes—look at those teeth! They started small in the center of the spiral and became progressively larger as they wound to the outer perimeter, where they were nearly as long as Troll's index finger. Finally there was the sheer mystery, the heavy curtain of time that no one had managed to draw back. Where did that buzz saw of teeth fit on the animal? How did it work? What sort of mega predator would have chompers like that?

Stewart was able to tell Troll and Matsen the genus of this whorl-toothed beast: *Helicoprion*. A smattering of such fossils had been found around the world. He could tell them that *Helicoprion* hunted the primeval oceans long before dinosaurs walked the earth—something like 270 to 280 million years ago, during the Permian period of the Paleozoic era. That was all Stewart, or anyone else, could say with conviction.

As the fluorescent lights buzzed and flashed, a monster shark materialized in Troll's imagination like a genie let loose from a bottle. When he and Matsen took their leave, a shadowy presence stalked him up the stairs and out the door.

———

Fast-forward seventeen years, to 2010. A scientific fuse was about to be lit, a stone rolled back in revelation. As these things are wont to do, it started serendipitously, in the basement of a museum.

Jesse Pruitt was in his third semester at Idaho State University student and working as an intern at the Idaho Museum of Natural History. Not the typical ISU undergrad, Pruitt was twenty-seven years old and a well-inked marine combat veteran. His first tattoos had come even before his enlistment, in big block letters spelling out the words "Hunt" and "Fish" across the backs of his fingers. Pruitt was born and raised in the tiny town of Golden, Mississippi, a boy's dream of rivers and creeks, lakes and woods, fish and deer and squirrels. He had a wiry red beard and blue eyes, was soft-spoken, quick to smile, and built like the proverbial brick shithouse.

For his internship, Pruitt was cataloging Ice Age mammals under the senior collections manager, Mary Thompson. One of the world's most fruitful repositories of fossils from the Pleistocene epoch (about 2.6 million to 11,700 or so years ago), was a mere twenty-five miles from campus. The site is second only to the tar pits of California, although excavation is complicated by the fact that it's mostly covered by a Bureau of Reclamation reservoir. As water levels rise and fall behind American Falls dam, however, it can be a veritable gumball machine of bits and parts of fossil camels, mammoths, bigheaded llamas, dire wolves, saber-toothed cats, giant sloths, long-horned bison, and short-faced bears.

Engineers made a sensible choice locating the dam where they did, at mile 714.7 of the Snake River. The water impoundment potential had been proven more than a million years before, when a lava dam only a few miles down from the present concrete dam formed a broad, shallow lake. Eventually the lava dam broke, draining the lake and leaving a basin filled with about eighty feet of sediment. Although the ancestor lake disappeared, the Snake River lived on, carving its channel through the gently rolling plain that the lake sediments had formed. The river drew animals to its banks to drink, and perchance to die, and more perchance to be fossilized, and way, way perchance to be found by people wearing hats and sunscreen kicking around on the muddy shores of American Falls Reservoir. Along with flooding who knows how many fossils, the concrete dam also drowned Shoshone and Bannock tribal bottom lands, part of the Oregon Trail, and the original townsite of American Falls.

Pruitt himself liked to go out to American Falls to look for stuff. He had always loved to be outdoors exploring, looking for treasures, and thinking about the ways nature worked. He was especially fascinated by dinosaurs as a kid—what they might have looked liked, how they lived and behaved, where they had gone. As he grew up, people told the bright, inquisitive boy that dinosaurs were not a job. You can't make a living doing that. What are you going to be, Indiana Jones? So in 1999, fresh out of high school, Pruitt joined the marines. He chose avionics as his specialty, which meant a minimum five-year commitment. First stop was boot camp at Parris

Island, South Carolina, followed by training at a base in Pensacola, Florida. He was eighteen years old, free of his small hometown and Baptist upbringing. There were girls, the beach life, a few fights. Pruitt washed out of avionics and went into diesel mechanics, learning the systemic intricacies and idiosyncrasies of Humvees and five-ton cargo trucks. Then on September 11, 2001, a jetliner hit the Pentagon, another crashed in a field in Pennsylvania, and two brought down the World Trade Center towers in New York City. Suddenly being in the military meant something very different from what it had the day before. In 2003, Pruitt was assigned to an artillery battery in a combat unit in Iraq, tasked with keeping the diesel trucks rolling, even during engagements. He was resourceful, steady, and absolutely kept the trucks rolling.

Pruitt left the military in 2004 with a fine-tuned skill set and a well-matched wife, Lena, a Montana woman and fellow military veteran. He and Lena landed in Mississippi, but before too long moved to Idaho Falls to escape the heat and bugs and to be closer to Lena's sister. Pruitt took a mechanic's job at a car dealership and resumed his aspirations of competing as an amateur strongman, an interest that began with powerlifting in the marines. He amped up his training, practicing Atlas stone lifts, tire flips, the farmer's walk, and other moves. Then the recession hit. The dealership laid him off, but Pruitt was able to pick up a job at a junkyard. In 2007, two weeks before his first official strongman competition, the van he was working on slipped off its jack and fell on him. The accident shredded his rotator cuff and demolished his strongman aspirations. An end, and a beginning.

After his shoulder injury, dark days of recovery stretched into tedious months of uncertainty. Lena urged him to think about college. He had never been much of a student growing up back in Mississippi, but he was smart and it would be a fresh chance. They could go together; Idaho State was just down the road. He had to do something. His first idea was to study paleontology, but ISU didn't have an undergraduate paleontology program and the Pruitts didn't have the resources to move. And besides (the voices crept in once more), you can't do that for a living. So he chose chemical engineering for the earning potential. But after a while he knew that money wasn't everything. You love what you love. Idaho State did have an undergraduate anthropology program, and he figured he could study early hominids, a subject he was interested in too. Pruitt went to see the museum's then director, Skip Loess, about an anthropology internship. After talking with Pruitt awhile, the director had a new message for the boy from Golden: go for it. You're meant for paleontology. We'll figure out the classes and cobble up a degree.

Pruitt crafted a double major in geology and biology and started his internship, sorting and cataloging American Falls material. As they worked together, Thompson saw Pruitt's potential for a career in science, and she mentioned that if he was considering going on to graduate school, it would be good to have done some undergraduate research. He had been mulling over her advice, thinking of what he could do for what he intended as, and assumed would be, a quick little research project, but nothing had leaped out at him. Hunters—whether a deer hunter or fossil hunter—have a sort of a third eye that is always processing signals from the surrounding physical world, regardless of where the hunter's conscious mind is. One October day in 2010, Pruitt walked back into the museum's work area to begin his day and noticed a big brown slab of rock on an orange flatbed dolly. He leaned in for a better look, brushing a hand over the cracked and slightly crumbling brown rock. A logarithmic spiral furled from the center of the slab like a giant fiddlehead fern. He saw teeth. Huge, pointed teeth, like Bowie knife blades, lining the curves. The whorl was about the size of a bicycle wheel, some two and a half feet in diameter. The toothy spiral conjured up a memory, matching an image already lodged in Pruitt's brain. This was that mysterious whorl-toothed shark. He never missed the Discovery Channel's Shark Week, and he remembered a segment on this freaky, mysterious, perfect predator. Bull's eye! Pruitt fixed *Helicoprion* in his sights.

2

THE SHARK BITES

Science is an intensely human endeavor.
—Keith Stewart Thomson,
The Common but Less Frequent Loon and Other Essays, 1993

MR. DAVIS, LIKELY WITH HIS SUPPLIES USED UP AND HIS BOOTS IN NEED OF RESTITCHING after all that prospecting, humped back to Perth. Western Australia's capital city was bustling in the 1880s, a noticeable shift from its start as an isolated agricultural settlement fifty years earlier. It's easy to imagine that after scrubbing off the last bit of red dirt at the municipal bath, Mr. Davis walked over to the pub for a meat pie, reading in the *Inquirer and Commercial News* over his beer that a government geologist had found traces of gold in the Kimberley. It would be up north to the Kimberley region for him, then. But today, as soon as he finished his pint he would look for someone who might be interested in that extraordinary, and possibly valuable, fossil—the mysterious talisman from his ride out of the bush.

Perhaps a miner loitering in the assayer's office suggested the name of Reverend J. G. Nicolay, a clergyman in the nearby harbor town of Fremantle who studied rocks and fossils and was said to welcome unusual finds. A suburban railway had just been put into service between Perth and Fremantle and the ten-mile ride would be a novelty in any case. So we can picture the freshly shaven Mr. Davis, with the world's first *Helicoprion* fossil in a cloth sack, catching the train to meet Reverend Nicolay.

The historic scientific community is a pyramid of individuals. At the top are the most famous minds: Copernicus, Darwin, Watson and Crick, Rachel Carson, and the select others in that esteemed intellectual royalty that are household names today. Buttressing the entire rest of the pyramid is a jostling throng of curious men and women from all walks of life out there observing, measuring, searching, pondering, collecting, writing, theorizing, debating, and sharing, mostly in relative anonymity—like Reverend Nicolay. We know slightly more about Nicolay than we do about Mr. Davis, but not by much. He could have been a Congregationalist, Wesleyan, Episcopalian, or Presbyterian minister, as each of those denominations was present in Fremantle at the time. Almost certainly he was involved with the small, local natural history society that would eventually morph into the Royal Society of Western Australia. Indeed, a few years after his meeting with Mr. Davis, Reverend Nicolay became a corresponding member of the already established Royal Society of South Australia, contributing a paper on the "Coast Limestones of Fremantle, Western Australia." It wasn't unusual for men of the cloth to involve themselves in the coalescing disciplines of geology and paleontology. After all, ordained theologians were a scholarly group. Charles Darwin himself almost

entered the clergy. And going back in time to the early natural philos-
ophers, God's word as written in scripture and his works as witnessed
in nature were viewed as mirroring aspects of the same truth. Far from
being in inherent conflict with religious faith, geology was accepted as
its accessory—evidence for the grandeur of God's creation.

Yet it was only a matter of time before the harmony struck a sour
note. And it was, literally, time that proved to be the most dissonant
chord. In 1650, James Ussher, Archbishop of Armagh and Primate of
All Ireland, worked some inspired math and determined that God created the world and every shark, spider, coal seam, Swan River blackbutt, and apple tree in it during a weeklong burst of celestial energy
that began at "the entrance of night preceding the 23rd day of October,
the year before Christ 4004." Ussher didn't explain what happened
to the first few weeks of October or what was just outside night's
entrance. It's easy to lampoon his divine date stamp, but Ussher was
a highly respected scholar, and the information didn't come to him
in a vision but through his industrious application of the most current knowledge of ancient history, astronomy, biblical languages,
chronology, and other disciplines. And it was no prayerful pondering
that led him to the rigorous academic work that ended with establishing
this date. As a Protestant bishop in a majority-Catholic land, Ussher
was obsessed with establishing the scholarly superiority of his reformed
faith over the resolutely intellectual Jesuits, an order of Roman Catholic priests. After setting the date of creation, Ussher further calculated
that receding floodwaters lowered Noah's ark upon Mount Ararat on
Wednesday, May 5, 2348 BC. The Great Flood as set forth in Genesis
(only six chapters after Adam and Eve are settled in the garden), was
regarded as the defining cataclysm that left the world in the shape that
Ussher beheld on his travels between Ireland and England.

Whispers that the world was vastly older rose like steam from a volcano not too long after Ussher's pronouncement. Aroused by such murmurings, the stripling discipline that would become geology began a long, slow branching away from the gnarled trunk of natural philosophy. Two revolutionary ideas had emerged by the mid-seventeenth century: that fossils were the remains of once-living animals, and that the rocks containing those fossils were formed in consecutive layers over some significant period of time. Far longer than what Ussher's date of creation would allow. This observation of the sequential nature of rock layers was profoundly important and gave us the first of two main organizing principles of modern geology: the Principle of Superposition.

It seems only fitting that sharks bearing coded messages from deep time would be the agents of this ripping quantum leap. In 1666, fishermen working the rich waters of the Ligurian Sea, an arm of the Mediterranean, caught what was likely a great white shark. Was it intentional? Or an unintentional heart-in-their-throats encounter? Did it charge their boat? Get tangled in their nets? We don't know, but it was a thrilling enough event that when the fishermen dragged the beast ashore word spread all the way to Ferdinando II de' Medici, Grand Duke of Tuscany. A patron of the sciences, Ferdinando ordered the shark's head be brought to his exceptionally bright young personal physician, a twenty-nine-year-old Dane with the Latinized name of Nicolaus Steno. This was a propitious collision of events. As Steno meticulously dissected the great white's teeth away from its gummy jaws, he was struck by their resemblance to *glossopetrae*, or "tongue stones," interesting little natural curiosities found embedded in certain rocks.

Tradition held that when Saint Paul was bitten by an adder while shipwrecked on Malta, he cursed the island's serpents, turning their tongues to stone. Tongue stones, especially those collected on Malta, were said to have powerful talismanic and medicinal properties and were often sewn into pockets or worn as pendants around the neck. (Thank Saint Paul, not a surfer, for your shark-tooth necklace.) At that time the word "fossil" was used for any natural curiosity dug from the ground, including crystals, archaeological artifacts, unusually shaped rocks, and those so-called tongue stones—as well as the prolific other stones that bore an uncanny resemblance to seashells. As various voices explained it from assorted angles, those objects had fallen from the sky, or were placed inside the rocks by the hand of God as a reminder of his creativity, or they grew inside the earth from a "hidden virtue," or had been formed by a shaping force (*vis plastica*) within the earth, or were simply sports of nature.

After Steno's close encounter with the great white, he added his voice to the mix with a new explanation for "fossils," outlining his ideas in a cautious appendix to a larger study on muscles that was published in 1667. Gathering his arguments and his convictions, Steno then wrote "Preliminary Dissertation Concerning a Solid Body Enclosed by Process of Nature Within a Solid," a bolder statement asserting that the tongue stones were the remains of once-living animals and suggesting how they came to be embedded in rock layers. This 1669 publication, commonly called the *Prodromus*, is considered by many to be the world's first geological treatise. In it, Steno drew a sharp distinction between his direct observations and his "conjectures." He built his potentially unsettling arguments with intention, from observation to inference to conclusion, pioneering a standard for rational scientific discussion based on empirical evidence. As proof that fossils weren't growing like virtuous but doomed potatoes inside the rocks, he noted that the *glossopetrae,* the tongue stones, appeared somewhat deteriorated, an indication that the rocks must have formed around them. He further argued that the stone enclosing them had once been soft, as suggested by the fact that they weren't distorted or deformed. That soft material was very probably wet, he thought, like the muddy sediments they all knew accumulated at the bottom of an abandoned quarry; the hardening of the layer must have occurred as moisture leached out. Educated people of the time understood the basics of dissolved minerals, and Steno proposed that those substances could make their way into organic remains (a carcass at the bottom of the quarry), transforming the original physical composition into a stony object. So, therefore, it was quite reasonable to believe that *glossopetrae* were the teeth of sharks that had died during a period of sedimentation.

It followed, Steno argued, that new layers of sediment would be laid down on top of older ones. And there it was, the Principle of Superposition: any layer, as originally laid down, is younger than the layer it rests on and older than the one above it.

The Principle of Superposition was a great start. Still, the layering was an observation of relative time rather than a calculation of overall age. Steno's layers could form in a matter of days or months, and thus comfortably fit into Archbishop Ussher's time frame and worldview.

Steno's contemporary, British scientist and inventor of the universal joint Robert Hooke, was also arguing that fossils weren't sports of nature. A brilliant if bilious polymath, Hooke devised an illuminated compound microscope, which he used to examine fossils, among many other things. In his 1665 book *Micrographica,* with its etchings of a flea, fly eyes, and the first cells, Hooke described structural similarities between fossil shells and the shells of living mollusks, and between petrified wood and living wood. This led him to share Steno's conclusion that fossils had organic origins. The idea of fossils as once-living animals was a lot to swallow, but at least there was a direct correlation between Steno's gleaming white shark teeth and fossil shark teeth—you had a dead shark, and an older dead shark. Hooke introduced a dicier angle into the picture by scrutinizing fossils like ammonites (those lovely extinct spiraled cephalopods) that could not be perfectly matched with any living animal. "There have been many other Species of Creatures in former Ages, of which we can find none at present," he wrote. The idea of extinction, theologically unacceptable in a divinely created world, was now a shadow at the door.

So Steno and Hooke gave us real fossils and superposition, the first principle of geology, nearly four hundred years ago. According to the Principle of Superposition, sedimentary rocks form in layers that accumulate over time, with the oldest layer at the bottom and the youngest layer on the top. It's really that simple, and it really was that much of a eureka moment back then in the late 1660s.

Of course extinction is tied up in the broader context of real, not relative, time, to which we turn next.

—————

In the way that shark teeth led Steno to fossils, fossils led English surveyor William "Strata" Smith to the second founding principle of modern geology: the Principle of Faunal Succession.

Credit came belatedly and grudgingly to Smith, born to an Oxfordshire blacksmith in 1769. Smith's father died when the boy was eight, and William was sent for a while to live with his farmer uncle. It was a time in history, writes Simon Winchester in his book, *The Map That Changed the World: William Smith and the Birth of Modern Geology*, when "the faiths and certainties of centuries past were being edged aside, and the world was being prepared, if gently and unknowingly, to receive the shocking news of scientific revelation." Revelation that Smith himself would drop on the world.

Young William, conceivably feeling dislodged and lonely away from his mother, began to fill his empty spaces with observations. The farm had a dairy and William noticed that his aunt—like all the dairymaids of the region—used a distinctive, dome-shaped stone as a counterweight on her balance scale when she was measuring out a package of butter. Locals knew them as pound stones, or Chedworth buns, for their happy similarity to plump hot cross buns. Chedworth was only about thirty miles away and the stones were abundant in area fields. It was easy to collect stones that were nicely uniform in size and weight—twenty-two ounces to be precise, or what the dairy managers called a "long pound." Hefting the smooth, comfortable weight of a pound stone in his palm, William noticed that it had five longitudinal sections (like a five-sectioned orange) and was incised with attractive repeating patterns. Maybe he spent parts of his days wandering the pastures, collecting the stones, comparing their markings before dropping them off at the creamery. The next curiosity to catch young Smith's eye were stones the size of small grapes, strewn through surrounding pastures just begging children to gather them up for a spontaneous game akin to marbles. They were most widely known as lampshells, for their resemblance to early Roman oil lamps, but the local name for those stones was "pundib." No explanation for the name seems to have survived, although it appears to have been an old miner's term. Juggling a few pundibs from hand to hand for the pleasant clack, William admired their varied textures and sheen.

Even as a boy, William must have intuitively known that the stones had a story far beyond marbles and butter, but it would have been hard to find anyone to tell it to him. Almost certainly not his uncle, or aunt, or his mother's new husband, or even his teacher at the village school, who was said to lead class with a cat sitting on each knee. But by then, the latter half of the eighteenth century, a small number of informed naturalists *did* know the story, or at least parts of it. They knew that Chedworth buns were fossilized sea urchins and that the pundibs were a type of fossil brachiopod. (Brachiopods look like clams to the casual observer but are in their own phylum, *Brachiopoda*. They're not even that closely related to mollusks, phylum *Mollusca*, which includes clams, snails, and squid.) Yes, Smith was born into a society that readily flocked behind the bellwether of theological doctrine. Yet it was also a time when new understandings were beginning to shear Ussher's wool away from open eyes.

When he was eighteen, Smith finagled a job as the assistant to a professional surveyor named Edward Webb, who taught him how to measure and value land. Smith impressed his new employer and was soon traveling the length of England and working independently. About five years after his hire, Smith was sent to make a survey of the Mearns Colliery, a working coal mine located on an estate sitting atop the Somerset Coalfield. Not wanting to (or not able to) spend money on coach fare, Smith walked the fifty miles to his new job along old Roman roads. Thus thoroughly grounded in Somerset geography we can see him on his first day of work at Mearns Pit, peering into a mine shaft, feeling the earth's own sultry breath on his face.

Spread across the Mearns Colliery were a number of shafts, and twenty-three-year-old Smith climbed down every one of them multiple times, sometimes descending by ladder, sometimes riding on a dredger chain. The dim underworld must have been heaven for Smith; his fascination with rocks had only grown from the pundib days, and he was eager to plumb earth's buried mysteries. First to draw his attention as he slipped deeper into the mines was the patterned layering of rock, with distinct and abrupt boundaries between each single layer, or stratum. Not too far below the surface he noticed orderly layers of red marlstone interspersed with greenish shales, like so many slices of bread and butter, he thought. He made an underlined mental note that the stacked layers dipped slightly downward, pointing in the general direction of London.

There are three main types of rock: sedimentary, igneous, and metamorphic. Simplistically speaking, sedimentary rock forms when sediments like silt, minerals, and decayed organic matter accumulate in horizontal layers that are eventually consolidated into solid rock. Shale, sandstone, and limestone are prime examples. Igneous rock, such as basalt and granite, comes from volcanic activity. Metamorphic rock is either igneous or sedimentary rock that has been transformed by tremendous heat and pressure: limestone becomes marble; mudstone becomes slate; granite typically becomes gneiss. Most of the rocks in our story are sedimentary, because that's where the fossil action is. The extreme temperatures and pressure that create igneous and metamorphic rock obliterate the possibility of fossils.

Underlying the red marls and green shales that Smith saw were more sedimentary layers, including sandstone, siltstone, mudstone, coal seams, and what is called "seat earth"—an ancient soil layer that supported the seed ferns, horsetails, club mosses, scale trees, and other plants that had transformed into the coal Smith and his employer were chasing. The miners knew each seam of the Somerset coal measures by name, like Dungy Drift, Kingswood Toad, Temple Cloud, Peacock, Warkey, Firestone, Globe, White Axen, and on. "Coal measure" is the term for coal-bearing layers of earth laid down between 310 and 290 million years ago during the Carboniferous period. The word root "carbon" is your big clue as to what that period was all about. Along with its given name, each coal seam had a distinct character. Blindfold a miner (if you dared), drop him down any of the collier's shafts, to any level in the coal measure, remove the blindfold, and he would know exactly what seam he was looking at. Each seam had its own thickness (anywhere from nine inches to nine feet), a distinct texture or shade, maybe a certain oily feel or scent—and very significantly, each rock layer in the measure had its own unique set of fossils. There might be freshwater bivalves like *Carbonicola* and *Anthracosia* in a layer of mudstone, marine brachiopods of the genus *Lingula* in a siltstone stratum, and ferny, palm-frondish *Mariopteris* and *Annularia* fossils in a seam of coal.

Smith wouldn't have run into any *Helicoprion* fossils there in the coal measures because those Carboniferous layers were older and deeper than the layers holding our beast. *Helicoprion* would be lying in layers some twenty to forty million years above Dungy Drift—with our old friend Marsh's *Apatosaurus* another hundred million years' worth of layers above that. Although Smith wouldn't have found *Helicoprion* in the Carboniferous measures, he might have run into other fossil sharks there, since the Carboniferous was a time of prolific abundance and wild evolutionary experimentation among the sharks-and-kin clan, which we'll explore in later chapters.

Moving up and down the Mearns Pit shafts, witnessing the predictable regularity in layering and fossil distribution, Smith began to formulate a monumentally important theory: that sedimentary rocks laid down at one particular time in one particular place will share the same key characteristics—most notably the same fossils—and will appear in the same stacked order, no matter where you find them. In his book on Smith, Simon Winchester calls Mearns Pit as significant to the science of geology as Gregor Mendel's pea garden was to genetics, the Galápagos Islands were to evolutionary theory, and the University of Chicago football stadium was to the story of nuclear fission. In the course of his other duties Smith predicted and confirmed his theory clambering down into mines miles apart and recording the same rocks in the same order with the same fossils. If this level of predictability occurred in the Somerset coal measures, could the same thing be true for all the rocks of England—or the world? Smith thought it could. He just needed a bigger sandbox to prove it. Or a canal . . .

Smith was working during a time when the smokestacks of the Industrial Revolution were belching with commerce, famished for coal. The cheapest way to transport that coal—in our story, Somerset coal—was by canal. Smith was an obvious choice to set such an excavation in motion, and in 1795, he was appointed the Somerset Coal Canal's first surveyor, putting him in the delicious position of being paid to lay open a great swath of countryside. As the machinery cut the earth like a cake knife, Smith saw that the rock layers were in the position he predicted, with each one containing "fossils peculiar to itself," as he knew it would. It was one of the most important breakthroughs in earth science and would usher in the modern discipline of geology.

In 1815, Smith published the first comprehensive geological map of Britain, covering England, Wales, and parts of Scotland. No doubt owing to his humble background and lack of social standing, Smith's discoveries were first ignored and then plagiarized by the scientific community, particularly the Geological Society of London. Formed in 1807, the Society was admirably the first body committed to the primacy of observation and fieldwork. Less admirably, the Society assumed itself to be, in addition to a scientific organization, an exclusive gentlemen's club whose members regarded geology as something a bit more elevated than Smith's practical, vocational application. The group's maltreatment of Smith and his work was nasty business and a devastating turn for the struggling freelancer. He began selling off his fossil collection but still ended up spending about ten weeks in debtors' prison, scratching out a living as an itinerant surveyor on his release. Seeing what was happening, a small dissident group of former colleagues and supporters mounted a campaign to give Smith the enormous credit he was due and pave his path to scientific redemption. Finally, in 1831, the Geological Society of London attempted to set things right by awarding William Smith the very first, very prestigious Wollaston Medal, geology's version of the Nobel Prize, or in Smith's case maybe the Purple Heart. (Smith never was invited to join the Society.)

Smith received the Wollaston Medal and the world received the Principle of Faunal Succession—the geological ground rule that unique fossil sets in rock layers succeed one another. The extinct clam *Carbonicola* will never, ever share a layer with the equally extinct but far younger *Helicoprion*, and we will never find the toe prints of dinosaurs and humans side by side.

The marriage of Steno's Principle of Superposition with Smith's Principle of Faunal Succession made it possible to classify rock layers and order them in relative time. Of course time is more than relative, and a giant, thorny question remained: How long had it all taken?

Bibles still carried the annotated date "Before Christ 4004" for Genesis chapter 1, verse 1, but there was a growing belief among naturalists that the earth must have been shaped and populated over an enormous span of time. By the close of the 1700s, murmurings of an older earth had become louder mutterings by bold and radical thinkers like Erasmus Darwin, the grandfather of Charles and a card-carrying "lunatick." (Lunaticks met monthly on the full moon to discuss ways in which scientific discovery might be applied to the mushrooming arena of machine-driven industry.) Certainly most people of the day consulted the Bible to answer life's eternal questions, but a growing number of free-thinking observers were asking important new questions: How did the earth *really* form? Over what time frame? Was there such a thing as extinction? What was the truth? What *was* truth?

Answers to these questions came from an unlikely sage—a Scottish farmer named James Hutton. Hutton's interest in chemistry led him to a medical degree in 1749, but he couldn't muster the enthusiasm to practice, so he went to run the family farm in the hills and lowlands of southeastern Scotland, near the North Sea. He was a progressive and engaged agriculturalist who forged his views on the forces of nature over fourteen years of feeling the wind and rain on his barns and his own back, and witnessing erosion and siltation, freezing and thawing, birth and death. He appears to have relished it, writing to a friend in 1753 that he had become "very fond of studying the surface of the earth," and was "looking with anxious curiosity into every pit or ditch or bed of a river" that fell in his way. Around 1768, when he was forty-two, he left the farm and returned to Edinburgh to focus on his scientific interests and puzzle over such things as fossilized seashells lying far above sea level. He read about the chemist Joseph Black's experiments with heated limestone, and saw the drastic power of heat as harnessed by the steam engines redefining British industry and life. Could the center of the earth be a massive heat source perpetuating processes that resulted in the endless destruction and reformation of rocks? He agreed with Steno's principle that sedimentary rocks were formed sequentially through deposition and consolidation. But where Steno thought the layer hardened through simple precipitation, Hutton believed the solidification of rocks was due to pressure and heat—the explanation that stands today.

In 1785 (when William Smith was a teenager thinking, it might be supposed, about dairymaids' Chedworth buns), Hutton presented a paper to the Royal Society of Edinburgh proposing that the earth is constantly being worn down and renewed by an endless cycle, with "no vestige of a beginning, no prospect of an end." The way Hutton told it, the world had a far longer, slower story than the one Archbishop Ussher told, which could be read more accurately in rock layers than between the lines in Genesis. Hutton published his two-volume *Theory of the Earth* in 1795, asserting that barely noticeable geological forces at work in the present day are identical to those that shaped the earth in the past. His theories formally introduced the jarring possibility that the earth was formed over an incomprehensible period of time. Hutton was clear in his exclusion of miraculous intervention as a force for earthly geological change. No mountains flung up in the course of an angry divine earthquake, or canyon scoured out on command. Hutton didn't deny that God created the universe. He simply considered it largely irrelevant to his inquiry.

Eventually Hutton would be recognized as the founder of modern geology, but certainly not right away. His ideas—as staggering and potentially disruptive as they were—spread at the speed of erosion, maybe because of his dense and difficult writing style. It took lawyer-turned-geologist Charles Lyell, another Scotsman, to roll the boulder down the hill fifty years later, around 1830, while scholar and wordsmith William Whewell coined the term "uniformitarianism" to capture the idea that planet-altering processes happen in a uniform way through time. Solidly rooted by Hutton's work, the Theory of Uniformitarianism was deployed as a full-throated counterargument against the more established Theory of Catastrophism, espoused by Lyell's own colorful and charismatic professor at Oxford, the theologian, geologist, and paleontologist William Buckland.

Buckland was a great, if quite eccentric, popularizer of science in the early nineteenth century. He determined to study his way through the animal kingdom by eating, sampling everything from toasted mice to roasted rhino. The biblical geologist was famously said to have snatched from a reliquary and eaten a morsel of King Louis XIV's heart, and to have licked purported saint's blood from the flagstones of a cathedral and pronounced it bat urine. Buckland brought us the word "coprolite" for fossilized feces and pioneered coprology, pushing the point by having a side table inlaid with polished coprolite sections. Buckland inaugurated the scientific investigation of "giant extinct lizards" in 1824 with an account of *Megalosaurus bucklandii*, almost twenty years before Richard Owen coined the term "dinosaur." Caught in the early-nineteenth-century wave of intellectual transition, Buckland identified himself as a "natural theologist" and strove to reconcile geology and religion, including Noah's Flood.

Lyell, who was among the most vigorous voices calling to disallow scriptural or theological influence in the science of geology, grew frustrated with Buckland's insistence on linking earth science with biblical revelation. Instead, Lyell wanted to finally and conclusively stand geology up as a true science based on evidence and observation. Buckland was following the lead of his friend and kindred spirit French anatomist Georges Cuvier, the leading scientific proponent of catastrophism. Cuvier's personal legacy is sticky; he was widely viewed as egotistical and vindictive, and was vehemently opposed to popularizing or democratizing scientific knowledge. (He also had a famously large head to match his conspicuously large ego.) In addition to all that, Cuvier was undoubtedly brilliant and exceedingly influential. He established the fact of extinction, formalized the methods of comparative anatomy, and almost single-handedly founded the discipline of vertebrate paleontology. But he was wrong about catastrophism. He painted himself into the corner of catastrophe for a reason: he rejected early Enlightenment notions of organic evolution (known then as "transmutation of the species") but embraced the truth of extinction, so he had to somehow explain the succession of fossil species now evident in the fossil record, thanks in large part to William Smith. For Cuvier that "somehow" came in the form of sudden, swift, violent, possibly global events in which some species went extinct while other popped into being in a system of progressive creation. Cuvier's argument against evolution was that species only breed with the same species, and an animal's anatomical makeup is so perfectly and delicately balanced that any meaningful change would disrupt that organism's ability to survive. Although he used scientific reason against evolution, Cuvier's religious convictions were deep-seated, and it was hard to fit evolution into an even allegorical reading of Genesis. God would give us only perfect creations. While he was the embodiment of reason in his scientific practice, Cuvier accepted religious truth as existing apart from reason. He did accept that the world could be several million years old, which would be time enough for species to go extinct and new ones be repopulated under the catastrophic plan while simultaneously denying evolution.

Uniformitarianism, on the other hand, required more than a few million years. Uniformitarianism unfolded along an almost unimaginably long time line, which Lyell estimated at more than three hundred million years, with mankind on the scene only very lately. As a former lawyer he knew how to shape a persuasive argument, which he did in his monumental *Principles of Geology*, published in three volumes from 1830 to 1833 to promote uniformitarianism's central argument that "the present is key to the past." Charles Darwin had a copy of Volume I aboard the HMS *Beagle*, and used it to interpret rocks and landforms he saw on the voyage. This idea that geologic change is the steady agglomeration of small changes over immense periods of time made a powerful impression on Darwin. After the voyage Lyell invited Darwin to dinner and the two became close friends. They often discussed their ideas, although it took Lyell a very long time to come around to an even lukewarm endorsement, in the tenth edition of *Principles*, of his friend's theory of evolution. Darwin, meanwhile, used Lyell's principles to form his own theory of the earth through evidence of uplift and subsidence seen along the coast of Chile and near coral reefs, especially in earthquake zones. This theory would eventually form the foundation of plate tectonics.

Lyell's tightly wrapped concept of uniformitarianism was overly neat, especially the notion that major changes always happen at the same dawdling pace, not accounting for extreme catastrophic events like asteroid strikes. Nevertheless, after the release of Lyell's *Principles* most geologists swung to the side of uniformitarianism—which they generally continued to do so all the way to 1980, when Walter and Luis Alvarez suggested an asteroid caused the great dinosaur die-off at the close of the Cretaceous. Today, most geologists agree that the earth's geological history is a long, slow, uniform symphony punctuated by a few crashing cymbals of catastrophe.

———————

By the time Mr. Davis knocked on Reverend Nicolay's door in the early 1880s, Charles Darwin had published *On the Origin of Species*, anesthesia and the phonograph had been invented, Louis Pasteur had developed a rabies vaccine, and the first issue of the journal *Nature* was on the stands, with articles on British moths, science education, and the Suez Canal. Our *Helicoprion* surfaced for the first time on a surging tide of scientific progress.

Let's see it as a sunny afternoon when the good reverend opened his door to greet his caller. Perhaps Nicolay asked about the train ride and offered tea in the study. There may have been niceties, but surely both men were eager to get down to the reason for the visit. Carefully removing the fossil from the sack, Mr. Davis placed it on the table between them. With a whistle, or a cluck, or a "my word!" Reverend Nicolay grabbed his magnifying glass for a closer inspection. He had never seen anything like the specimen lying before him. No one had, ever, other than Mr. Davis and the mule. The men must have enthusiastically passed the glass back and forth, speculating on the animal that once wielded those wicked blades in its quest for survival eons ago. Mr. Davis told Reverend Nicolay where he had collected the fossil, which Nicolay recognized as a place where rivers cut through a flat-topped range, composed of limestone overlaying concretionary sandstone. The reverend might have quizzed Mr. Davis on what else he had noticed, possibly asking about *Spirifers,* brachiopod fossils he knew to be common in the area. Nicolay might have reached for his own box of the elegantly winged shell fossils to show his interesting guest.

Reverend Nicolay may or may not have paid Mr. Davis for his exceedingly rare treasure, but even if he did, their exchange was more than a mere financial transaction. The meeting was collegial enough that when Reverend Nicolay later sent the fossil on, he attached a specific request that it be named for Mr. Davis.

Reverend Nicolay hadn't had the unidentified fossil in his possession long when the new governor of Western Australia, Frederick Napier Broome, and more important to our story, his wife, Lady Barker, steamed into the Fremantle harbor after a long, roundabout journey from Broome's latest post in the Republic of Mauritius. From her letters, it sounds as though all Lady Barker wanted to do was bail out of the storm-tossed boat and drag herself the last fourteen miles to their new residence at the Government House in Perth by any means other than a boat. But when the boat docked, a large crowd was there to greet the couple with champagne and enormous bouquets of flowers. Lady Barker, around fifty years old at the time, rallied her most gracious composure. This was Australia's winter, Perth's high social season, which the Broomes quickly realized they were expected to host. Three weeks later they were throwing parties and greeting guests by the mansion's grand staircase. One early visitor wrote in his diary that Lady Barker "has evidently laid herself out to be agreeable to people and has made herself acquainted with the peculiarities of some people."

Lady Barker had some notable peculiarities herself. Described by the same visitor as "a fine tall woman, with well-marked features and a somewhat decided manner," she was born Mary Anne Stewart in 1831 in Spanish Town, Jamaica, where her father served as island secretary. When she was twenty-one she married Captain George Robert Barker, bearing him two sons and following him to India. Captain Barker was knighted in 1859 for his service during the bloody Indian Rebellion, and Mary Anne became Lady Barker. Captain Barker enjoyed his status only a short time; he died in Bengal eight months after the knighting. In 1865, Lady Barker kept her title but married Frederick Napier Broome, a strapping sportsman, poet, journalist, and diplomat eleven years her junior. Broome persuaded her to leave her sons to their education in England and go with him to New Zealand to operate a sheep ranch that he had purchased from Richard Knight, Jane Austen's grandnephew. Lady Barker relished her time in New Zealand but the brutal winter of 1867 killed half their sheep, putting an abrupt end to their ranching days. Broome then took a series of diplomatic postings from Natal to Trinidad, while Lady Barker wrote books, including the popular *Station Life in New Zealand*, in which she describes New Zealand's flora and fauna and recounts traveling on horseback, eel fishing, pig stalking, hunting wild cattle, and skating. "I am afraid that it does not sound a very orderly and feminine occupation, but I enjoy it thoroughly," she wrote, "and have covered myself with glory and honour by my powers of walking all day."

Reverend Nicolay could have been in the champagne-and-flowers crowd that greeted the new governor and his well-connected wife in

Fremantle, or he might have been a guest at one of their parties in Perth. Maybe he had read Lady Barker's books. It seems entirely plausible that over libations the two would have fallen into a conversation about Australia's peculiar nature and Nicolay's fascination with rocks and fossils. How could he not tell his vivacious, interested hostess about the very unusual specimen that this character Mr. Davis had carried out from the Gascoyne watershed? We can almost hear her. "Oh, Reverend, how very fascinating! My brother-in-law Robert Scott is a Fellow of the Royal Society in London. He loves that sort of thing. Why don't you give me a letter describing your curious fossil, and I will send it on to him. Maybe he can tell you something about it."

Indeed, on July 13, 1883, Reverend Nicolay wrote a letter describing the fossil, attaching a photograph likely made with George Eastman's new dry-plate process. Lady Barker sent the correspondence on, and when the letter and photo arrived in her brother-in-law's post halfway around the world, it was like our shark bit him on the leg. Scott replied to his sister at once, wanting to see this fossil in the flesh, so to speak. Could she possibly arrange it? Of course she could. In a flurry of notes and letters, Lady Barker enlisted the help of Edward T. Hardman, an Irishman who had been serving as Western Australia's government geologist. It was Hardman's report of gold in the Kimberleys that Mr. Davis might have seen in the newspaper. Hardman's position had been defunded and Lady Barker caught him as he was about to regretfully catch a steamer back to Ireland.

Now safely stowed in Hardman's luggage, Mr. Davis's fossilized furl of teeth rested in its stone cradle as the ship crossed an ocean salted with hammerhead sharks and lobe-finned coelacanths and myriad other denizens of the deep with close and direct family ties going back millions of years. But were those fossil *Helicoprion* teeth to suddenly reanimate and sprout jaws, gills, fins, and tail, and slip overboard, the reconstituted beast would find itself completely alone, without relatives among the living fish families.

Hardman handed off the fossil to Scott in 1885 on his way through London. (While wanting nothing more than to return to Australia, Hardman would die of typhoid fever in Dublin the following year.) Scott knew immediately that he was looking at something special. He sent the letter and photograph—but not the fossil—to his Royal Society colleague and curator of the British Museum's geological department, Henry B. Woodward. Upon seeing the photograph of the fossil, Woodward entreated Scott to deliver the specimen to him without delay for his own examination. And at last the mysterious fossil came to rest near the center of the universe for nineteenth-century earth sciences, the office of Henry B. Woodward. It had been quite a journey for the random chunk of broken rock, which, apart from the inexorable drift of tectonic plates, probably had not moved more than a few feet in more than two hundred million years.

3

RIGHT SHARK, WRONG NAME

The greatest obstacle to discovery is not ignorance—it is the illusion of knowledge.
—Daniel J. Boorstin, 12th Librarian of Congress, 1975-1987

WHEN THE PORTLY AND WELL-SETTLED HENRY BOLINGBROKE WOODWARD TOOK possession of the strange specimen from Western Australia around 1884, he was one of the world's foremost fossilists and editor of the esteemed *Geological Magazine*. Woodward's path might have been predestined, but it wasn't easy. Born the son and grandson of geologists, baby Henry came into the world in 1832, between Volumes I and III of Charles Lyell's *Principles of Geology*. Henry's father, Samuel Woodward, died when Henry was six, leaving the boy to be raised in large part by his geologist elder brother, Samuel Pickworth Woodward. Henry supported himself as a bank clerk until he was twenty-six, when he earned an assistant's position at the British Museum. Six years later, in 1864, he cofounded *Geological Magazine* and would serve as its sole editor for fifty-three years. (Still published by Cambridge University

Press, the periodical remains one of the most prestigious earth sciences journals today.) The genial Woodward was said to be a judicious and tactful editor who welcomed honest work, "whether on orthodox lines or otherwise." He shaped the journal to be a hub for sharing information and a venue for lively debate, often with American scientists who were publishing in their own proliferating journals, including the *Proceedings of the Academy of Natural Sciences of Philadelphia*, *Scientific American,* and the *American Naturalist* (all still being published today).

Although Woodward's specialty was fossil crustaceans, he wrote widely on many topics, including the first feathered fossil *Archaeopteryx*, his thoughts on possible fossil links between birds and reptiles, and the geological importance of beavers. Two of his daughters were illustrators and frequently supplied drawings for his articles, as well as for other paleontologists. Gertrude Mary Woodward grew into a prominent scientific illustrator in high demand, and her sister Alice B. Woodward was one of the most prolific illustrators of the time, creating exquisite illustrations for such classics as *Peter Pan* and *Black Beauty*, in addition to her scientific work. In 1868, Henry Woodward wrote about the first Ice Age mammoth found in England, describing the "true inward curvature" of the tusks and pointing out that Russian scientists had illustrated the tusks going the wrong way in their article about the Siberian mammoth. If Woodward showed brass in his writings we might forgive him. He was making a great career for himself, he was liked by his peers, and his sails were tight with the strong winds of nineteenth-century scientific advance.

Library shelves were becoming populated with monographs on fossil invertebrates as well as on the dinosaur and other fossil vertebrates stampeding out of the ground in Marsh and Cope's raging Bone Wars over in America. The focused study of fossil fish was slowly gaining momentum too, emerging from the shadows of the then flashier vertebrate fossils. Just a few miles away from the British Museum where Woodward eagerly awaited the package from Robert Scott, London's Natural History Museum was organizing its extensive new fossil fish collection, founded largely on specimens accumulated by two Oxford-educated gentlemen, William Willoughby Cole (3rd Earl of Enniskillen) and Sir Philip de Malpas Grey Egerton. Egerton's interest in paleontology was fanned by the Oxford lectures of our quirky coprologist William Buckland. While traveling together in Switzerland as young men, Egerton and Cole met Swiss-born geologist Louis Agassiz and determined to study fossil fish. Between them, the two men gradually assembled the largest and finest private fossil fish collections in the world. Agassiz, who emigrated to America in 1847 to teach at Harvard, would be regarded by history as a manipulative polygenist autocrat, but his scientific contributions were enormous and indisputable. His publication *Recherches sur les poissons fossiles* ("Research on the fossil fish"), released in five beautifully illustrated volumes from 1833 to 1844, launched the field of paleoichthyology and expanded the number of named extinct fish species to more than seventeen hundred.

It was good times for paleontology and for Woodward. By the time he received the Australian fossil from Robert Scott he had been in charge of *Geological Magazine* for some twenty years and was head curator of the British Museum's geology department. No wonder he felt so sure of himself. Solely from the photograph that Scott had sent ahead, Woodward had already begun formulating his opinion on the Australian fossil. On the day he received the specimen itself, he might have carried it to the window to more closely scrutinize the fan of serrated points he'd been examining in the black-and-white image. The British Museum had installed electric lamps six years earlier in some parts of the museum, one of the first public buildings in London to do so, but one would want to shed the brightest light available on such inscrutable traces of the past.

Finally arranging himself at his desk, Woodward overconfidently dipped pen to ink and wrote that, even before he had the physical specimen in hand, "I readily identified the fossil photographed as the impression of a fish-spine, similar in form, but more highly curved than those . . . originally described by Prof. Leidy as a fish-jaw, and named by him *Edestus vorax* in 1855."

By lumping the Australian fossil in with the *Edestus* genus, named thirty years before by American naturalist extraordinaire Joseph Leidy, and by believing the treacherous barbs to be fin spines rather than teeth, Woodward had just joined what would eventually prove to be the wrong camp. In the final flourish of his monograph Woodward wrote, "I have been requested to append the discoverer's name to this fossil, and as I am unwilling, in the present state of our knowledge, to make a new genus for its reception, I propose to name it *Edestus Davisii*."

It would be fourteen more years before a Russian geologist would finally, correctly, create a brand-new genus for Mr. Davis's fossil: *Helicoprion*.

———————

A quick catch-up on taxonomy is in order here, because in some ways it is the warp and weft of the *Helicoprion* story. Taxonomy is nothing more (and so much more) than the naming of things.

The accurate labeling of life-forms is critical to scientific integrity, even though names can be a moving target, as we'll see. History, circumstance, and scientific politics can all influence which names are ultimately assigned to what organisms.

The foundations of modern taxonomy were set by Carl Linnaeus, born in Sweden in 1707 to a Lutheran pastor who cultivated a flourishing rectory garden. Little Carl loved nature and would ask the names of plants, only to promptly forget them. Finally depleted of Lutheran grace, his father said there would be no more new names until the boy started remembering the ones he'd already been given. So the seed for Linnaeus's eventual work to simplify scientific nomenclature was likely planted as he sat among the buzzing bees and fragrant blossoms straining to recall yet another mind-numbing name.

Naturalists in those days were already using a loose Latin-based system of genus and species to identify plants and animals. But the names were often overly long and arbitrary. With the deluge of unfamiliar plants and animals from around the world arriving at European ports in the dank ship holds of adventuring sea captains like Vitus Bering, James Cook, George Vancouver, and Alexander von Humboldt, a more consistent system of classification was sorely needed. In his game-changing book *Systema Naturae,* published when he was twenty-eight years old, Linnaeus proposed using a two-word identifier for each organism, in a sort of last-name, first-name arrangement. The Latin- or Greek-based "last name" would designate the genus ("race or stock") and the "first name" would designate the species ("specific or particular kind"). The species name, also derived from Latin or Greek, would reference something relevant to the organism, like a color, memorable characteristic, geographic location of first discovery, or discoverer (like our *davisii*). You can almost hear the sighs of relief over this codified simplification, along with inevitable grumblings over change and status quo.

With genus and species as the sort of handshake introduction to a plant or animal, Linnaeus created broader groupings of kingdom, class, and order. Later scientists added more tiers, including phylum, which classifies organisms based on their body plans. For example animals in the phylum Arthropoda—the arthropods, which encompass everything from trilobites to tarantulas to king crabs—have segmented bodies, jointed appendages, and an external ("exo") skeleton. The phylum Chordata, which contains all the vertebrates including whale sharks and humans, possess at some stage of life gill-like slits, a tail, and a notochord, which is the sort of starter backbone that gives vertebrates our head-to-tail (mouth-to-anus) body plan. Even though we humans only have gills and tail as embryos, those embryonic clues are important in determining evolutionary relationships. The more narrowly defined tier of "family" was also added to Linnaeus's original set. Family is directly above genus, like Canidae for the canine family or Hominidae for the human family.

Today's taxonomic system has settled in with eight levels. From broadest to most specific they are domain, kingdom, phylum, class, order, family, genus, species. It's not necessary to remember them, but a stew pot of mnemonic devices are available if you do want to upload the list to your big vertebrate brain. One of the most common is Dear King Phillip Came Over For Good Soup. Less common but more memorable: Didn't Know Popeye's Chicken Offered Free Gizzard Strips. Oh, those biology students.

———————

What we know now that Henry Woodward didn't know then, sitting with Mr. Davis's fossil at his desk in London, is that there was a very old, very weird, dead-end group of sharks that had either curved (as in the *Edestus* genus) or fully spiraled (as in the *Helicoprion* genus) tooth structures positioned in the center of their mouths. You need to know about *Edestus* (aka the "scissor-tooth shark") because it took a while for scientists to parse *Helicoprion,* with its single tooth whorl, from *Edestus,* with its paired, curved tooth blades, and even when they did finally sort them out, they often talked about them in the same breath—even though they were notably different fish from significantly different time periods. Still, it's an understandable impulse to lump them both together, since the oddly curved tooth structures made them wholly unlike all the other fish from the chondrichthyan clan.

You don't have to memorize the taxonomic hierarchy, but if you don't already know the word "chondrichthyan" it will be a useful term to remember. All fish—most famously sharks—that have skeletons made of cartilage rather than bone belong to the class Chondrichthyes. This is the big tent for all the cartilaginous fish. You can never go wrong using the word "chondrichthyan" whether you're talking about a great white shark, a *Helicoprion,* a stingray, or a spotted ratfish.

Eventually, a whole new order would be erected for the chondrichthyans with the curved and spiraled tooth bases. To prepare the way for fully appreciating the swarming seas of prehistoric sharks, let's think about how inclusive and diverse the taxonomic categories of class and order can be. Since our own lineage is most familiar, we'll find our illustrations there. Humans are in the class Mammalia, which hugely contains humans, bats, mice, and every other furry, suckling creature. To make the giant class a little more manageable, Mammalia is divided into three subclasses: the placental mammals (subclass Eutheria;

manatees and humans for example), the marsupials (subclass Metatheria; such as kangaroos, koalas, opossums, and the extinct Tasmanian wolf), and the monotremes (subclass Prototheria; the platypus and anteaters). Within the placental subclass of Eutheria, humans are in the order Primates, relegated there with the lemurs and gorillas by our shortened snouts, clever hands, and collarbones, among other things. One more breakdown because it's sort of a joke on us and also shows the tendency of systemicists to erect pup tents inside larger taxonomic tents, the order Primates is divided into two suborders, the Strepsirrhini—or primates with moist noses—and the Haplorhini, the primates with dry noses. Humans are in the group with dry, though sometimes runny, noses.

The point is: a class is enormous, orders are big, and the sorting funnels down from there. The scientific classification of life-forms is only hard at first. Learning any new language, including this one, can be fun and satisfying once you get a feel for the lingo.

Back to fish. The class Chondrichthyes is divided into two sub-classes, generally agreed to be the Elasmobranchii and Holocephali. General agreement is often the best you can hope for in the study of fossil fish. The field is in steady flux, based on the latest finds and most recent technologies and methods. Elasmobranchs have usurped the title as "real" sharks, like the megalodons and makos, while holocephalans are interchangeably called chimaeras, ghost sharks, and ratfish. Strictly speaking, you're not supposed to call a holocephalan a shark. That seems fair today when the only holo-cephalans still living fit into one, single order, the Chimaeriformes, which don't look or act at all like sharks. You would be right to wonder why such a great big division—a whole subclass!—has been set aside for one single order. The answer is, because most of the chondrichthyan action happened a very long time ago, and the subclass Holocephali was once plugged to the gills with orders, families, genera, and species, some of which looked very sharklike and others of which were nothing less than finned hallucinations. Since we are not taxonomists or natural history museum staff members, we are going to do ourselves a favor and follow the "gestalt" rule. Any ancient chondrichthyan that looks like a shark, with impressive teeth, a torpedo-shaped body, and sharky fins and tail, we can call a shark for our purposes. This talk about holocephalans is a clue as to what lies ahead for our decidedly sharklike beast *Helicoprion*.

Imagine you are the first person to see a fossilized animal part that looks somewhat familiar yet strangely unique, all at once. In 1855, that's what happened to Joseph Leidy, the extraordinary, all-but-forgotten figure who named the brand-new genus *Edestus,* to which Henry Woodward mistakenly added Mr. Davis's Australian fossil. With Leidy, our story touches down on American soil.

Joseph Leidy was born in Philadelphia in 1823. His father, Philip, worked in the family hat-manufacturing business. Twenty months after Joseph was born, his mother, Catherine, died giving birth to her fourth (second surviving) child. A few years after that, Philip married Catherine's cousin, who loved the dreamy Joseph as fully and fiercely as she loved the six other children, who followed the wedding. Young Leidy was a mediocre student but ardently pursued his personal interests. Enthralled by nature, he routinely skipped school to roam the banks of the Schuylkill River and Wissahickon Creek, Bartram's Garden (a botanical garden and arboretum), and the other natural areas

Philadelphia still had in its back pockets. According to Leonard Warren, author of *Joseph Leidy: The Last Man Who Knew Everything*, Joseph was so incorrigibly truant from school and late to dinner that his parents hired a "responsible companion" to keep track of time and protect the bookish boy from local ruffians. Leidy's tall, broad-shouldered roaming buddy was an African American youth named Cyrus Burris, son of the Leidy family's laundress. Burris knew the surrounding countryside well from his work collecting herbs for a local physician, and sometimes the boys would stay out all day fueled by nothing but raw turnips scrounged from the fields. Around the time the preoccupied youths were mucking about in local marshes, Darwin was setting sail aboard the HMS *Beagle* with Lyell's *Volume I* under his bunk.

From childhood, Leidy had a genuine gift for scientific illustration. When he was ten he filled a sketchbook with sixty-five shell drawings, annotated with their scientific and common names. His father had no use for intellectual people and wanted Joseph to leverage all that artsy nonsense into a vocation as a sign painter. But with his mother's intervention and encouragement Leidy entered into a medical apprenticeship with a private anatomy teacher, later enrolling at the University of Pennsylvania medical school. After graduating with a medical degree in 1844, the passionate naturalist began a halfhearted practice. About a year later he took on a side project dissecting and illustrating snails for a book on North American mollusks, and his exemplary work brought recognition and membership into both the Boston Society of Natural History and Philadelphia's Academy of Natural Sciences. This turn of events was all Leidy needed to abandon his medical practice, although he parlayed his medical school education into a lifelong career as a professor of human anatomy at the University of Pennsylvania. Teaching bankrolled his true love, the broadest possible study of nature. Leidy didn't just look at anatomy as his meal ticket, he was thoroughly fascinated by it, and his exceptional grasp of comparative anatomy became his key to the paleontological kingdom.

Comparative anatomy, as a discipline formalized by Georges Cuvier, is essentially the comparison of similarities and differences in lifeforms. For instance, comparative anatomy tells us that we humans are related to lobe-finned fish but not to ray-finned fish. How does it tell us that? If we lay out the skeletons of a human, a coelacanth, and a perch and compare them side by side, we see that the coelacanth and the human skeletons both have humerus bones, but the ray-finned fish does not. Granted the humerus of the human and the coelacanth are shaped quite differently, but they serve similar skeletal functions, as a major bone in the forelimb helping connect shoulder to hand/lobe-fin. That makes the humerus a "homologous" structure (*homo*, same) that both we and the coelacanth inherited from a common ancestor. Every bone in our body corresponds to a "matching" bone in every other vertebrate, though remodeled by evolution or shrunken into disuse like the vestigial leg bones in snakes. Whales have a pelvic girdle; we have a pelvic girdle. This is "divergent" evolution, where two species that are very different now, like us and the coelacanth, come from a common ancestor. Contrast that to "convergent" evolution, where species have evolved similar features or behaviors, like wings and flight, even though they don't share a common ancestor. Bats and birds are an example. The wings of bats and birds are "analogous," not "homologous." The ability to ferret out the biological relationships of extinct animals is what elevates paleontology from a name game to an evolutionary understanding of ancient life.

Leidy typically worked until 2:00 A.M. every night and most Sundays, a schedule that allowed him to develop his authoritative expertise in anthropology, paleontology, entomology, botany, zoology, geology, mineralogy, pathology, parasitism, and microscopy. He was one of the last of the great generalist scholars. Even after astonishing his colleagues by marrying at age forty-one, Leidy's happy union to Anna Harden easily accommodated his unfettered curiosity and intellectual ambitions. Anna had a keen interest in biology and assisted her husband with the microscope and drawings.

Among all of Leidy's interests paleontology rose to the top, and he made enormous contributions to the emerging field. It was Leidy who described the first American dinosaurs *(Deinodon, Trachodon, Troodon,* and *Palaeoscincus)* from teeth collected along the Missouri River in present-day Montana. In the years before the Bone Wars, Leidy was the acknowledged leader of American paleontology. Collectors—including his friends Ferdinand Hayden of the US Geological Survey and Spencer Baird of the Smithsonian—sent him fossils from all over the country. He was like a spider in the middle of a great collecting web, with the most interesting finds of the day landing at his feet like fat flies.

If you do happen to know the name Joseph Leidy, it's probably in association with the duck-billed dinosaur, *Hadrosaurus.* In 1858, Leidy's friend William Parker Foulke, an abolitionist, prison reformer, lawyer, and fellow member of the Academy of Natural Sciences, was vacationing in Haddonfield, New Jersey. On social rounds Foulke met a prominent local landowner named John Estaugh Hopkins, who showed Foulke some giant bones that had been found twenty years earlier in a marl pit on his farm. (Marl, a lime-rich mud or mudstone, was mined extensively in New Jersey in the 1800s as a soil conditioner.) Foulke was eager to look for more of the bones, and convinced Leidy to oversee the excavation and interpretation. Between the bones Hopkins already had and follow-up excavations, about 30 percent of the skeleton was recovered—producing the most complete dinosaur skeleton in the world at that time. Leidy gave it a new genus and species, *Hadrosaurus foulkii*. He described *H. foulkii* in 1860, although publication was delayed by the Civil War. During the war Leidy served as assistant surgeon and pathologist at Philadelphia's Satterlee General Hospital, which treated thousands of Union soldiers and Confederate prisoners. Besides the tragic humanitarian toll taken by the Civil War, American science was also set back by the conflict, halting research, interrupting debate about Darwinism and evolution, disrupting communication with European scientists, and dislocating scientific organizations. However Leidy leaped back into his work after the war, releasing ten papers in 1865, including the one on *H. foulki,* as well as papers on fossil horses and rhinoceroses, fetal dog sharks, human bones found in a guano deposit, and Cretaceous sponges and reptiles.

In 1868, Leidy worked on a reconstruction of the hadrosaur with sculptor Benjamin Waterhouse Hawkins, who had done Sir Richard Owen's models for the Crystal Palace dinosaurs in London. Until then, real dinosaur bones had been displayed in cases as isolated specimens. Under Leidy's guidance, Hawkins suspended real and plaster bones on a metal armature, stood the thirty-foot herbivore up on its back legs, and gave the world its first dinosaur skeleton display. When the hadrosaur was unveiled in the hall of the Philadelphia Academy of Science, a gasp arose on both sides of the Atlantic. Where Owen's dinosaurs were squat quadrupeds walking on all fours, Leidy's hadrosaur sported a kangaroo-like stance, based on its revealed skeletal structure of small forelimbs and large rear leg bones—a posture that paved the way for Godzilla and *T. rex* to stomp into our collective cultural consciousness. The hadrosaur was so popular that visitors overwhelmed the museum's volunteer staff and raised enough dust that curators protested. The academy started charging an admission of ten cents and opened their doors only two days a week until the facility could be moved into a new building.

For this and his other accomplishments Leidy is considered the father of American paleontology, even though he was subsumed in popular memory by the rapacious Bone Wars. The paleontological pirateers Othniel C. Marsh and Edward Drinker Cope, the latter of whom studied anatomy under Leidy, treated the elder scientific statesman like a relic. They ignored his contributions and grabbed up all the vertebrate bones for themselves, leaving Leidy to fade into obscurity and return to his old interests in parasitology and protozoology. For his behavior, Cope became the only person the sweet-tempered Leidy ever detested.

Over his lifetime Leidy published more than eight hundred scientific articles. In most of them, he stuck to straight description and known fact. But in a paper published six years before the 1859 publication of Darwin's *On the Origin of Species,* he uncharacteristically mused on the continuum of life. "There appears to be but trifling steps," he wrote, "from the oscillating particle of organic matter to a Bacterium . . . and so gradually up the highest orders of life." As soon as he read *On the Origin of Species,* Leidy wrote to Darwin saying, "I felt I had groped about in darkness, and that all of a sudden, a meteor flashed upon the skies." Like other open-minded scientists, Leidy had been grasping for the mechanism through which new species arose, which Darwin provided in his ideas about natural selection.

"Your note has pleased me more than you can readily believe," Darwin wrote back, "for I have during a long time, heard all good judges speak of your paleontological labors in terms of the highest respect. Most Paleontologists (with some few good exceptions) entirely despise my work. . . . All the older Geologists (with the one exception of Lyell whom I look at as a host in himself) are even more vehement against the modification of species than are even the Paleontologists." Leidy, who normally avoided controversy, was among the first to step up and publicly support Darwin's ideas, advocating for Darwin's induction into the Academy of Natural Sciences of Philadelphia, Darwin's first official recognition following the publication of *Origin*.

These details are relevant to *Edestus* in that it establishes Leidy as a super-naturalist of unquestioned skill and repute. According to his biographer, "In paleontology, if Leidy didn't know, no one knew." With his deep knowledge across the spheres of natural history, Leidy had an uncanny ability to visualize and describe life-forms from fragments—a tooth here or a bone there—operating intuitively "with few or no guiding landmarks, formulating relationships and establishing new genera."

This was the talent he called to the fore when he received a very odd fossil fragment roughly the size and shape of a russet potato, with broken stumps of teeth. A mineralogist friend of Leidy's, William S. Vaux, came by the specimen from an itinerant showman, who said it came from "Frozen Rock, Arkansas River, twenty miles below Fort Gibson in the Indian Territory," which was the Oklahoma Territory. Fort Gibson had been established in the 1820s for the terrible work of Indian "relocation," and was the final, desperate destination for the Cherokee people after their 1838–1839 forced march on the Trail of Tears. In the years following the Indian resettlements and before the

Civil War—when Confederate soldiers briefly occupied the fort—
Fort Gibson tumbled into serious decay, with the surrounding lands
dotted by cheap saloons and brothels. Those doggeries were surely
on the circuit of the traveling medicine shows common to that
time, with their flea circuses, magic acts, and freaks. There could
have been such an establishment near Frozen Rock, which by the
mid-nineteenth century was a busy landing for steamboats traveling
the Arkansas River. A Confederate soldier who camped at Frozen
Rock described an exposure of shale there, where water running
from a seep froze into a sheet of ice in the winter. There's a good
possibility that the fossil Vaux gave Leidy wasn't actually collected
at Frozen Rock, because the jet-black specimen looked unweath-
ered, and its crevices were filled with "carbonaceous matter," as if
it had been taken from a coal mine. Later geologists would suggest
it came from Illinois. Perhaps a teetering fur trader approached the
itinerant showman with the curiosity, willing to swap it for a bottle
of laudanum. The showman gladly agreed to the trade, confident
he could use the strange piece as an entertaining attractant for cus-
tomers he could then charm into buying his questionable tonics.

*Come one, come all! Gather 'round. Not so close! Behold the night-
mare rippers of a devil monster that prowled among the giants of
Genesis . . .*

4

FIRST, COUSINS

Show me your teeth and I will tell you who you are.
—Georges Cuvier, paleontologist and comparative anatomist (1769–1832)

MR. DAVIS'S *HELICOPRION* FOSSIL WAS STILL LYING IN THE AUSTRALIAN BUSH WAITING TO be discovered when in 1855, Joseph Leidy introduced *Edestus vorax* to the world at a meeting of the Academy of Natural Sciences in Philadelphia. Along with his presentation, Leidy submitted a seventeen-line summary for publication in that organization's 1855 *Proceedings,* to establish naming precedence in advance of his deeper study and more detailed description of the fossil.

The first line of the entry read: *"Edestus vorax,* Leidy.—A species of a new genus of fishes founded on the fragment of jaw with portions of four teeth." Leidy didn't explain how he arrived at the name, but *edeste* is a Greek word meaning "devour," and *vorax* is Latin for "voracious." Apt enough. Judging by the tooth parts left in the fossil, he thought the intact teeth would have been about two inches long and two inches wide, probably resembling "those of *Charcharodon*." *Charcharodon* was an alternate spelling of *Carcharodon*, as in *Carcharodon mega-lodon*, the colossal fossil predator that Agassiz had identified and named in 1835. *Edestus vorax*'s coarsely serrated teeth weren't as big as megalodon's, but they would have still covered the palm of a child's hand. If his fossil had been whole, rather than the potato-size fragment it was, it would have been about a foot long.

The most puzzling aspect of the fossil to Leidy was the segmen-tation of what he presumed to be the jaw, which was formed in sandwiched wedges that spooned tightly together. Each wedge cul-minated in a tooth. From its general form, Leidy thought the fossil looked like part of a lower jaw—except for the fact that "no verte-brated animal, neither living nor extinct, has yet been discovered in which [the lower jaw is] segmented." He could think of a few fish with segmented upper jaws however, including Lepisosteidae, the ancient-but-surviving *gar* family, and a Devonian lobe-finned fish. "Therefore it is a fair inference that the fossil in question is a portion of an upper jaw."

So, dealt this wild card of a fossil by an itinerant showman from the Indian Territory, Leidy opens the historical hand with what seems like a good bet. Four teeth, upper jaw, giant shark, *Edestus*.

———

Then, where Leidy's *Edestus* had been the only one in the world, a similar fossil but much better specimen showed up that very year at a Providence, Rhode Island, meeting of the American Association for the Advancement of Science, in the possession of Edward Hitch-cock. Hitchcock had just stepped down from nine successful years as president of Amherst College and was settling into his new position at the institution as professor of natural theology and geology. If Leidy's fossil looked like a rough potato, Hitchcock's handsomely intact fossil looked like a large banana studded with seven fulsome teeth. Hitchcock's fossil had been found in a layer of shale overlying a coal seam in Parke County, Indiana, where a local doctor gave it to a local pastor, who sent it along to Hitchcock. Hitchcock's first instinct was that the fossil was "the jaw of a shark, but of very peculiar character." Louis Agassiz himself, then a Harvard professor, was at the meeting, and here is where the long story of "the blind men and the elephant" began for our curve-toothed sharks. In that Indian parable, one man touches an elephant's side and says the animal is like a wall. Another fingers the tusk and asserts the elephant is like a spear, while another seizes the squirming trunk and judges it to be like a snake.

After examining the new *Edestus* specimen, Agassiz declared that the studded spike must have projected from the head of a shark, like the rostrum (beak, bill, snout) of the living sawfish, *Pristis*. There must have been a matched set of them on the fish's head, he said, to act as a powerful offensive weapon. For the unique segmentation, Agassiz offered that the *Pristis* sword is comprised of two bones, and that if those bones should separate, it might look like the fossil under consideration. Agassiz was ready to create not only a new genus but also a whole new family for the specimen. He was apparently supposed to get it for deeper examination and naming, but for unknown reasons the fossil went instead to the regaled Sir Richard Owen at the British Museum. Owen assigned the fossil to the *Edestus* genus but asserted that the structure was a dorsal (topside) fin spine that must have been used as a defensive weapon. In his drawing of the fossil Owen distorted its shape to support his opinion, and even contemporaries who agreed with his opinion grumbled that he gave "a bad figure of it." Nevertheless, the fin spine theory was now firmly stuck in the literature.

Leidy had a bit of a waffle after he heard Owen's declaration. In the 1856 *Proceedings of the Academy of Natural Sciences of Philadelphia,* Leidy published a short entry saying, "Since describing the fossil, supposed to be the fragment of an upper jaw of a fish, to which the name of *Edestus vorax* was given, it has occurred to me that it may perhaps be the portion of a dorsal spine of a huge cartilaginous fish." He notes that he was informed by a colleague, who had seen the fossil for himself, that it did indeed look like an "ichthyodorulite," a term used for fossil fin spines. But then it's as if Leidy just can't fight his own impeccable instincts. He goes on to say the "form of the teeth and their relative position to one another . . . are the same in both fossils." *Teeth once again!*

It must have largely been a move to stay in the conversation, because three years later, when Leidy published his more detailed description, he had regained his bearings and described the specimen as the "fossil fragment of the jaw of a remarkable and gigantic fish." He still found the segmentation to be the most arresting thing about the fossil. "The most remarkable peculiarity of the jaw is its segmented character," he wrote. Each segment created a "bandlike" surface on the side of the jaw. More curiously, the base of each segment bent at nearly a right angle to nest under its neighboring tooth. "A careless inspection of the fossil," he wrote, "would mislead one to suppose the teeth were inserted by long fangs into the jaw." He further observed that the whole fossil appeared to be unified from the tip of the teeth to the base of the jaw. In fact, decades later paleontologists would finally come to understand that *Edestus's* "jaw" was actually a rigid root—so Leidy's observation of the unified structure, if not his use of the word "jaw," was correct.

Leidy made another key observation about *Edestus* teeth. At first glance they did look like megalodon teeth, but closer examination revealed a number of unusual differences—foremost that the teeth appeared to be symmetrical, with no obvious inside and outside surface. Put your tongue to your front teeth and you will be reminded that the inside surface is cupped, while the outside is a little bulged. Human or shark, it's usually not very hard to tell the side facing in ("lingual") from the side facing out ("buccal" or "labial"). But *Edestus*'s teeth showed no difference. This important detail sheds light on why scientists clung so hard to the idea that the fossil was a fin spine rather than teeth. Even more than Leidy's lump of a fossil, Hitchcock's well-preserved *Edestus* was a head-scratcher. It didn't follow the bowed shape of a jaw, and if it were split lengthwise down the middle it would be bilaterally symmetrical. (In bilateral symmetry, something can be divided along a vertical plane into right and left mirror images.) Just as there was no precedent for a segmented lower jaw, there was no precedent in nature for a bilaterally symmetrical tooth structure; nothing to nudge the imagination toward a single line of teeth in the middle of the jaw. It was much easier to make sense of the structure as ridging along a fish's back than as a jaw.

Discussions of *Edestus* quieted for a while, since there was nothing new to talk about. Then about ten years after Leidy and Hitchcock presented their respective fossils at their respective meetings, the distinguished geologist, physician, explorer, and fossil fish authority John Strong Newberry came into possession of a new fossil from Posey County, Indiana, for which he erected a second *Edestus* species, *E. minor*. A few years after that, Newberry and his associate, Illinois state geologist A. H. Worthen, received a fossil from John P. Heinrich, that had been found in Heinrich's Belleville, Illinois, coal mine. Newberry and Worthen felt this new fossil was different enough for its own species, so they named it *E. heinrichi*. Not too much later Newberry and Worthen described yet another new species, the whopping *E. giganteus*, based on a fossil from the coal measures of Macon County, Illinois. This one was as big as a boomerang, with three-and-a-half-inch . . . spines? Teeth?

Newberry, who might have had the most impressively long and bushy beard of all the splendid beards of nineteenth-century paleontology, supported Owen's assertion that the *Edestus* fossil represented a defensive fin spine. This opinion had heft coming from Newberry, who was a very accomplished, highly regarded, and well-liked figure. The youngest of nine children, Newberry was two years old when his father moved the family in 1824 to a parcel in the former Western Reserve at Cuyahoga Falls, Ohio, near the sleepy village of Cleveland. The entrepreneurial Newberry senior began mining coal while the junior Newberry immersed himself in Ohio's wildlands, methodically cataloging the abundant plants of the Cuyahoga Valley and collecting piles of fossils, including many out of his father's mines. Most were plant fossils, but there were fish too. Newberry grew up but he never grew out of his interests, even after graduating from Cleveland Medical School in 1848. Following a two-year sabbatical in Paris with his new wife, Newberry continued to study the natural sciences while establishing a medical practice in Cleveland. In 1855, he was made a United States Army physician and appointed to serve as the surgeon-geologist-botanist on an expedition to explore the area between San Francisco Bay and the Columbia River. It was the end of his private medical practice and the beginning of a life of scientific adventure and achievement. Between 1857 and 1859, Newberry served as physician and naturalist on three major expeditions to the wildest parts of the western territories, including an exploration up the Colorado River "as far as its great cañons," making him very likely the first geologist to see the Grand Canyon. At the outbreak of the Civil War Newberry reported to the War Department in Washington, DC, and was appointed secretary of the western department of the United States Sanitary Commission. From the division's headquarters in the Union stronghold of Louisville, Kentucky, he supervised the Mississippi Valley region, a major front in the war. (In a postwar accounting, Newberry reported that his department had distributed, among other supplies, about 100,000 bottles of whiskey and "other stimulants," nearly 40,000 ounces of chloroform, more than 2,000 pounds of disinfecting powders, 317 ounces of quinine, 137 pounds of blackberry root, 56 pounds of slippery elm bark, 19 pounds of camphor, and 12 boxes of assorted medicines.) A few years after the war ended, he settled into a twenty-six-year career as professor of geology and paleontology at the School of Mines of Columbia University in New York. Over his life Newberry helped found the National Academy of Sciences, was the president of the American Association for the Advancement of Science, and served as an officer in the Geological Society of America.

He was a prolific writer, notably on the subject of fossil fish. In 1870, Newberry and Worthen coauthored a detailed description of that third new *Edestus* species, *E. heinrichi*, which looked very similar to *E. minor*, but with slightly less overall curve and squatter teeth. Of course Newberry and Worthen didn't call them "teeth" but "denticles," which are basically modified scales. Chondrichthyan scales are very different from the scales of bony fish like salmon. Interchangeably called "placoid scales" or "dermal denticles," these elegant little structures are basically small, anatomically correct teeth, with a pulp core capped by hard, calcified tissue known as dentin, which in turn is mantled by another layer of enamel-like material. Don't picture a tooth-shaped structure though; think more along the lines of a streamlined stealth bomber, with leading and trailing edges, ridges, and a sleek chevron shape. Most typically, denticles occur as tiny shingles covering a shark's skin like chain mail—tough as nails, strong as steel, and flexible. In addition to protecting the skin, the textured denticles create little vortices that reduce drag as the shark swims, adding to its efficiency and, yes, stealth. On certain species, both living and extinct, denticles can also form larger spiky ornaments that look like thorns or even cat claws. The earliest identifiable chondrichthyan fossils were denticles, 440 million or more years old, with actual shark teeth finally appearing in the fossil record some twenty-five million years or so later.

Newberry and Worthen were talking about the thorny kind of denticles in their articulate 1870 description of *E. heinrichi* in *Paleontology of Illinois*. Their foot-long *heinrichi* had seven well-preserved "denticles" about an inch high, with sharp cutting edges and coarse serrations. The fossil absolutely belonged in Leidy's *Edestus* genus, they said, citing the odd segmentation. Then with utmost respect for Leidy's opinions, they politely reeled back his idea that the fossil was a jaw fragment. The mistake was understandable, they said, since Leidy's fossil was "exceedingly imperfect." Acknowledging his "proverbial acuteness and knowledge of comparative anatomy," they granted that "no other conclusion was fairly deducible from the fragment which he had." And yet, their fossil "exhibits features that seem to be incompatible with that conclusion." What now?

Newberry laid out his reasoned arguments that the fossil was a fin spine in his epic *Paleozoic Fishes of North America*, in the chapter "The Structure and Relations of *Edestus*." First, sharks do not have symmetrical teeth, and the *Edestus* structures were symmetrical. Second, all sharks by definition have cartilaginous jaws that are poorly fossilized, and teeth are only attached by ligaments, "so that in the fossil state the jaws have usually quite disappeared, the teeth being scattered about in all directions." *Edestus*'s toothy denticles, on the other hand, were firmly attached to a bony base. Third, "The form of this fossil, as shown by the nearly complete specimen before us, is wholly unlike that of any jaw of fish, reptile or mammal known." As if that wasn't enough right there, fourth: the rough, rounded base couldn't possibly have attached to bones or even cartilage, leaving it to most closely resemble the dorsal spines of sharks and skates which are implanted in the skin. It wasn't such a wild idea if you could ignore the toothy gestalt of those so-called denticles. After all, other fossil sharks had fin spines, like the six-foot-long *Hybodus*, which sported bony spikes in front of its dorsal fins, and the two-foot-long *Ctenacanthus*, which had distinctive, textured spines at the leading edge of its dorsal fin.

In the end Newberry's conclusion was exactly right and exactly wrong. The right part: "We are, therefore, driven by this perfect bilateral symmetry to suppose this was not one of a pair, but that it stood alone, somewhere in the medial line." The wrong part: that they were spines "used for attack and defense," and therefore must be located somewhere outside the body. For Newberry, it was simply too large a leap to imagine that any animal could have a symmetrical structure like that in its mouth. Where would it fit? How would it work? No animal they knew had any such thing.

New interpretations began to arise when Miss Fanny Rysam Mulford Hitchcock spoke up in 1887. And at last we can introduce a formally trained woman of science to our story. Make no mistake, women were already playing a valuable, if nearly invisible, role as research associates, illustrators, and collectors working with their husbands, fathers, and brothers—women like Anna Leidy, the important fossilist Mary Anning (the *she* in "she sells sea shells . . ."), the much-commissioned Alice and Gertrude Woodward, and Edward Hitchcock's wife, Orra White Hitchcock, one of the first well-known female botanical and scientific illustrators in the United States. With Fanny Hitchcock, however, we have a full-fledged woman scientist in her own right. (Fanny Hitchcock and Edward Hitchcock don't appear to have been related.)

Born in New York City in 1851, Hitchcock entered the University of Pennsylvania in 1890 as an undergraduate biology student at the mature age of thirty-nine. She was already highly educated, probably through some sort of private tutoring or European schooling. It's likely she enrolled at Penn to study under Leidy, because when he died the year after she enrolled, she immediately transferred to the Graduate School of Arts and Sciences. In 1894, Hitchcock became the first woman to receive a PhD in chemistry from the University of Pennsylvania. Following postgraduate work at the University of Berlin she returned to Philadelphia, becoming Penn's first director of women students in 1898 and establishing a women's athletic program. Her tenure, along with the rented women's gym that she paid for herself, was short-lived. In 1901, Hitchcock stepped down when university trustees rejected her proposal to create new opportunities for women's education—"with much regret," they assured her, "and a sincere sympathy in her desire to advance the education of women." From then on she maintained a well-equipped lab in her Philadelphia home, continuing to do research, write and publish papers, and encourage and financially support students who wouldn't otherwise have had the resources to pursue a higher education.

Hitchcock was a member of the American Association for the Advancement of Science, as was Anna Leidy. The AAAS was an inclusive organization from its early days, granting membership to women. It's a nice thought to believe that Fanny Hitchcock and Anna Leidy might have known each other. Anna was twenty years older, but she appeared to be socially engaged and had her own authentic interest in biology beyond wifely support. Joseph Leidy often credited Anna for her help with the microscope and illustrations, and she found some important fossils on their trip to Wyoming in 1872. Anna and Fanny both lived in Philadelphia, and they would have been sisters of a very small sorority of scientifically accomplished women. Anna lived until 1913, more than twenty years longer than Joseph, so maybe the two women sat together at occasional AAAS meetings.

Hitchcock was thirty-six years old and still living in New York when she presented her prescient paper on *Edestus* at an 1887 meeting of the AAAS—before she was even enrolled at Penn. First Hitchcock summed up the current thinking. There was theory number one, "Doctor Leidy's view, now no longer held," that the *Edestus* fossils were part of the upper jaw of an extinct chondrichthyan. Theory two was Richard Owen's suggestion that the fossils were fish spines, a view "also proposed by Doctor Newberry and . . . generally accepted until lately." And theory three, "the view recently proposed by Dr. H. Woodward of

London, in his account of a remarkable fossil from Australia," that the fossils were pectoral (side) fins. Woodward had drawn a parallel to the fossil fish *Pelecopterus*, described by Edward Drinker Cope, in which the pectoral fin was essentially a powerful spine.

Hitchcock then proposed her own original theory: that the *Edestus* fossils were teeth that grew in a center line between the jaws. She supported her ideas in five points, often drawing perspicacious distinctions between the American and Australian fossils, and describing why the existing theories seemed "doubtful" to "untenable." Bingo. Hitchcock was the first observer to call the fossils a midline tooth structure, and the first to call out major differences between the American and Australian fossils. As for the segmented nature of the fossil as well as the presence of a bony support, Hitchcock offered the example of *Onychodus sigmoides*, an extinct lobe-finned fish with very odd whorls of tusk-like teeth in the lower jaw.

Newberry was again the one to reel in the idea, politely, as he had done with Leidy, and at times almost reluctantly. He praised her as "an earnest and accomplished student of comparative anatomy" and conceded that, taken by itself, Hitchcock's idea was quite plausible. "There are perhaps no facts which disprove this hypothesis," Newberry writes, "and it is worthy of respectful consideration, but I would suggest that *Onychodus* was very widely separated zoologically from *Edestus*."

Certainly Newberry was correct on that point. *Onychodus* was a lobe-finned fish in the class Sarcopterygii, and *Edestus* was a shark in the class Chondrichthyes. That's a wide evolutionary gulf. Still, when Leidy offered the gar (class Actinopterygii) as his example of another fish with a segmented jaw, no one thought he was suggesting *Edestus* was a ray-finned fish. Maybe Newberry was grasping at straws. He agrees that the American and Australian fossils were very different, especially in their degree of curvature. But the blind men were stuck. "Wherever that species [*E. davisii*] goes," he pointed out, "*E. minor, E. heinrichi,* and *E. giganteus* must follow." Newberry could imagine *davisii*'s more tightly curled teeth fitting in the mouth of a ten-foot-long shark, but it was "scarcely comprehensible" to visualize the *giganteus* or even *heinrichi* fossils as part of a jaw of any sort. "Certainly such a monster would seem very much out of place in the lagoons of the coal marshes . . . If, now, we transfer this spine to the position of the post-dorsal fin, and bury it in the soft parts, all except the denticles . . . [this] becomes intelligible and natural."Newberry moved to close the discussion for the time being. "Until further light shall be thrown upon the interesting question of the homologies and functions of *Edestus*, we may regard them as the post-dorsal spines of large cartilaginous fishes of which the other parts are yet unknown, and may suppose that they were used for attack and defense."Fortunately that new light, and the Russians, were coming.

5

WHORL OF FORTUNE

He who calls what has vanished back again into being,
enjoys a bliss like that of creating.
—Charles Lyell, *Principles of Geology,* 1830,
quoting historian Barthold Georg Niebuhr

MR. DAVIS'S FOSSIL FRAGMENT, MISUNDERSTOOD AND UNRECOGNIZED, HAD WAITED
some fifteen years for a complete whorl to emerge and more fully tell
its story. Finally, here it was, a spellbinding new fossil that answered
one key question and raised myriad more that would flummox paleontologists into the twenty-first century.

What flukes of circumstance, proclivity, and luck brought the world
its first, fully spiraled *Helicoprion* whorl? With only a name, occupation, place—and the fossil—handed over to history, we are left once
again to construct a story out of scattered bits and pieces. It's not too
different from what paleontologists and artists do when reconstructing
incomplete forms: take what you know and add what makes sense.
Then sometimes go a step further and trek over the bridge of imagination to bump around in the world of possibilities.

It was probably the late winter or early spring of 1898. Maybe February. We can see the Russian school inspector, Mr. A. Bessonov, as he stepped from his sleigh and hurried toward the schoolhouse, with its thin plume of coal smoke rising from the chimney. At least the frigid temperatures left the roads passable. Bessonov would have been one of about sixteen school inspectors for the entire Perm district, a region larger than Iceland in the northeast portion of European Russia that spilled across the west slope of the Ural Mountains. Bessonov himself was likely responsible for more than a hundred schools scattered over almost eight thousand square miles. That day he was inspecting a school in the town of Krasnoufimsk, located in the central Urals about a hundred miles from his office in the city of Perm, the regional center. As the horse clipped along on its studded shoes, Bessonov might have pondered Russia's fragile state of affairs. The Romanovs were still in power and Nicholas II was on the throne, but a dissident underground was heating and rising like magma beneath Russia's aristocratic crust. The geographically isolated Perm province was far from the country's political centers so Bessonov probably wouldn't have caught the rumors, but even as he stepped through the schoolhouse door shaking snow off his coat, a small group of activists was planning a secret meeting in Minsk to form the Russian Social Democratic Labour Party. The young Marxist Vladimir Ilyich Ulyanov, who went by the name Vladimir Lenin, was in exile in Siberia, but he would smuggle the group draft articles written partly in milk between the lines of a book. Ulyanov/Lenin's own father had been a school inspector—reportedly one of the good ones, more interested in reforming the dire state of Russian education than in imposing an authoritative, sometimes inquisitional personal will, as other inspectors were sometimes inclined to do. In nineteenth-century Russia, school inspectors oversaw classroom routine, health conditions, and sometimes the teachers' personal lives. We have no idea where Bessonov fell on the bureaucratic continuum, but let's say he did the best he could and teachers didn't consider him a threat. In any case, they only saw him once a year on a hasty and perfunctory visit.

At the Krasnoufimsk school, Bessonov suggested to the school administrator that the students wash their hair more often, and had a teacher select two boys to stand and read a few lines. The days were short and Bessonov was anxious to get going, so he excused himself from the children's song and was leaving when a teacher pulled him aside to ask if he would let her keep her job if she married. He agreed to consider it. As Bessonov shrugged on his fur-lined coat, the thankful woman mentioned that one of her students recently brought in some interesting fossils from the town's limestone quarry, Divya Gora. Bessonov had once told the teacher that he collected rocks and curiosities, so she thought he might be interested. The report intrigued him.

Steam-powered rock drills hadn't made their way to Divya Gora yet, so the quarrymen were using pickaxes, sledgehammers, and chisels to break stone from the quarry walls when Bessonov stopped in on his way back to Perm. It was good construction-grade limestone, thick light-gray rock with a yellowish tint. Perhaps Bessonov knew that the stone falling under the pickaxes could have once been an ancient marine reef. The quarry manager said he had set aside some fossils just the week before, two rocks holding unusual spirals about the size of dinner plates, plus three related fragments. He had no use for them himself and would be happy to load them into Bessonov's sleigh.

Back in Perm, the overworked, underpaid inspector was swamped with paperwork. His annual report was due and he had caught a cold on the trip to Krasnoufimsk that he couldn't shake, but neither could he shake the idea that the Divya Gora fossils might be important. They were clearly different from the ammonites he collected in his youth. *Could those be teeth? Incredible!* Shoving aside his papers, Bessonov prevailed upon a journalist friend to take some photographs of the fossils, which he sent by post along with a letter to the Imperial Russian Geological Survey in Saint Petersburg. Bessonov also crated up the specimens and had them freighted the eleven hundred miles to Saint Petersburg.

The letter and photographs arrived first, landing on the desk of Alexander Karpinsky, distinguished Russian geologist and director of the Geological Survey. For the fifty-one-year-old Karpinsky, it must have been like getting a letter from home. He had been born in the Ural Mountains, in a copper-mining settlement where his father and both grandfathers were mining engineers. The Urals are one of the earth's oldest surviving mountain ranges, with the tallest peak worn to a hikeable 6,214 feet. Running north-south from the Arctic coast nearly to the Caspian Sea, the Urals bisect western Russia and form a natural demarcation between Europe to the west and Asia to the east. The mountains lie mostly in Russia, with the southern reaches extending into northwestern Kazakhstan.

Encyclopedic information like this is useful of course, but mountain ranges are more than their coordinates, far more than the sum of their parts—their peaks, valleys, rivers, steppes, forests, bears, wolves, and wildflowers. Mountain ranges shape a region's personality, cultural identity, and historical and economic destiny. This holds deeply true for the Urals. Buried under the Uralian geography is an ancient seam of mythology, beginning with the Paleolithic nomads who taught themselves how to keep wild beehives there tens of thousands of years ago. The Bashkir people, who have occupied the Urals since at least the ninth century, celebrated a hero named Ural-Batyr, a kind-hearted, honest, empathetic man of unflinching courage and great physical strength whose ultimate quest was to defeat Death itself. On his crusade, Ural-Batyr met giants and strange creatures, saved people from harm, tamed a wild bull, rode a winged white stallion, and married a swan maid. In the end, Ural-Batyr conquered Death by sacrificing his own life. Instead of drinking magic water he had won through a series of trials and hardships, he poured it on the ground to make Nature immortal. The grieving, grateful people piled stones over Ural-Batyr's body, and the burial mound become the Ural Mountain range, with the hero's body transforming into precious metals and gems.However it happened, the Urals do indeed hold bountiful riches. There are the fossil treasures, of course. Not just *Helicoprion,* but vestiges going back to life's earliest earthly history. Alongside those priceless treasures of immortal nature, the Urals contain some fifty useful ores and minerals, including gold, platinum, nickel, iron, lead, copper, coal, salt, topaz, emeralds, amethysts, aquamarines, sapphires, diamonds, and the rare and beautiful bluish-to-plum-colored alexandrite, discovered in the Urals and named for Tsar Alexander II. Peter Carl Fabergé, jeweler to the Russian court, relied on an Imperial lapidary factory in the central Urals to polish up Ural-Batyr's miraculous relics for his dazzling jewelry and famous eggs.

Some of the oldest mine sites in the Urals are copper mines dating back to the Bronze Age (bronze is a copper alloy), so maybe young Alexander Karpinsky kicked the fields around the mine where his father worked looking for artifacts. He wouldn't have had much time to build his boyhood collection though, as his father died when he was around eleven and he was sent to the Saint Petersburg Mining Cadet

Corps. He would go on to graduate at about the age of twenty with a degree in mining engineering from the Mining Institute (now the National Mineral Resources University, one of the oldest technical colleges in the Western world). A gifted geologist, Karpinsky's career path didn't lead back to the mountain copper mines of his father and grandfathers but to the marble halls of academia and science. Among his many endeavors, he taught at the Mining Institute for nearly three decades, where he was greatly respected and appreciated by his students. He considered his life a ministry of Truth in science, and was said to never refuse scientific help to anyone. His research and ideas became known worldwide and influenced the development of geology at the turn of the century, including early thinking about the movement of landforms. Despite his many obligations and involvements, Karpinsky ventured into his beloved Urals nearly every year to conduct fieldwork.

Karpinsky began his career with a focus in petrology, the origin, composition, and structure of rocks. But by the time he received the box of fossils from Bessonov, his attention had shifted to paleontology and paleogeography—what the earth's ecology was like in the geological past, and what forces had shaped it. To understand the earth's geological history one must know its extinct creatures, and Karpinsky especially relished "problematic" fossils, taking a great interest in the identification of mysterious forms. Young researchers often brought him specimens from the field they couldn't identify, both as their last resort and most trusted source. And now, here was this very strange whorl, most likely of Permian age, from the Perm district school inspector.

There is a connection between the Russian province of Perm and the Permian period—which makes this is as good a time as any to dive into the epic drama of the geologic time scale, with all its creative character development and sensational plot twists.

As there are acts in a play, you can think of earth's history as having four major acts. From oldest to youngest they are the Precambrian time, Paleozoic era, Mesozoic era, and Cenozoic era. Within each act are "scenes," or periods.

The Precambrian time was by far the longest act, covering about 90 percent of earth's history, from its superheated beginnings to about 600 mya ("mya" = million years ago). During this time the earth was sorting itself out from a molten mass into a differentiated planet with a core, a crust, and standing water. The Precambrian held life's prologue in an assortment of bacteria, and near the end of its run, an "Ediacaran biota" of tubular and frond-shaped organisms.

Earth's life story really picked up in the Paleozoic era (542–251 mya). The curtain rose on the Paleozoic with the emergence of complex life, and fell 291 million years later with the largest mass extinction earth has so far known. Fittingly, the very eventful Paleozoic is the second longest act, with six periods. Oldest to youngest they are the Cambrian, Ordovician, Silurian, Devonian, Carboniferous, and Permian. *Edestus* had its cameo appearance during the Carboniferous. *Helicoprion*'s story takes place in the Permian. Spoiler alert: *Helicoprion* was not a victim of the end-Permian mass extinction but winked out twenty million years or so before that calamitous event. Because so many important things happened across the long span of the Paleozoic—including the emergence of sharks—let's run through each period's high points.

The Cambrian period (542–488.3 mya) is life's Genesis chapter; its watery garden. You may have heard the term "Cambrian Explosion" to describe this scene, because this is when complex life arose, proliferated, and spun off into vastly diversified forms. Hard-shelled invertebrates evolved in the Cambrian, including straight-shelled cephalopods and trilobites. Cephalopods still live on earth, trilobites do not. And so it goes. The vertebrate ancestors also evolved in this period, opening the first door for both sharks and we humans. Eyes evolved during the Cambrian, as did the strategy of predation. No doubt the two were related.

The Ordovician period (488.3–443.7 mya) came next. Cephalopods added spiraled shells to their repertoire and figured out jet propulsion. They grew to enormous sizes of five feet and more, and staked a claim as the dominant predators of the period, hunting bottom-crawling creatures "like owls snatching mice." Fish made their rather timid entrance in the Ordovician, in the form of gilled, jawless creatures up to about twelve inches long. Most of the earth's landmasses clumped together in the Ordovician to form the supercontinent Gondwana. Two other smaller landmasses, Laurentia and Baltica, were also out there, adrift in the global Panthalassic Ocean. The inexorable clumping and unclumping of landmasses over the history of the earth through the forces of plate tectonics has driven evolution and extinction through its crashing, banging, rifting, uplifting influence. The global configuration of landmasses and ocean basins drives weather patterns. Weather affects environment, which shapes habitat, which establishes what life-forms—from sharks to dragonflies—can survive where, which triggers the underlying imperative for all living organisms: adapt, migrate, or die. Even before Gondwana there was the supercontinent of Rodinia ("Motherland" in Russian), the billion-year-old remnants of which can be seen in rocky outcrops in New York's Central Park.

The Silurian period (443.7–416 mya) followed the Ordovician. Gondwana rotated toward the south, glaciers melted, sea levels rose.

Vast coral reefs thrived in warm, sunlit waters. Jawless fish were the most common vertebrate, but with great evolutionary hubris, some Silurian fish developed a new superpower—jaws, arguably the most important twist in the vertebrate story. Jaws, and the teeth that eventually came with them, opened a whole new range of behaviors and survival strategies. Jaws could grasp wriggling objects, and teeth could rip the wrigglers apart to eat. Jaws could hold on to a partner during copulation (sharks do it, as do horned lizards), and jaws improved respiration by making it easy to pump oxygenated water over gills. Strong jaws and grinding teeth also allowed the full exploitation of seaweeds as food. And with jaws, who needed hands? Jaws empowered fish to move objects like small rocks, to get at what was underneath or to arrange nests.

While Silurian fauna were busy working their jaws, the flora stepped up their program too. Vascular plants evolved, with tissues (xylem and phloem) to conduct water and minerals, enabling them to begin colonizing all that empty Gondwanan real estate. With the plants came insects, including spider and centipede ancestors.

The Devonian period (416–359.2 mya) followed the Silurian. The Devonian was the great Age of Fishes. In a swelling chorus of jaws, all the fish groups were now onstage: the armored placoderms (class Placodermi), the cartilaginous fish (class Chondrichthyes), and the bony fish (superclass Osteichthyes), which included the ray-finned fishes (class Actinopterygii), and the lobe-finned fishes (class Sarcopterygii). Placoderms dominated Devonian seas, most famously the nightmarish *Dunkleosteus,* which grew as big as a school bus. With guillotine-like tooth plates and the bite force of an angry alligator, *Dunkleosteus* could slice a shark in two. Because now in the Devonian, there were sharks to slice. *Cladoselache* was one of the earliest. The streamlined six-foot-long predator (or prey, if *Dunkleosteus* was in the vicinity) had gill slits, two dorsal fins, and a distinctly sharklike snout and tail. The two-foot-long *Ctenacanthus* also arose in the Devonian. That's the shark with the cylindrical dorsal fin spines that the beautifully bearded J. S. Newberry used to support his argument that the *Edestus* fossil was a fin spine.

Incredibly, paleontologists found the fossilized body of a shark in Devonian-aged rocks. Discovered in New Brunswick, Canada, in 1997 and described in 2003, *Doliodus problematicus* is the oldest articulated shark fossil known. Articulated fossils preserve an animal's component parts in their life positions and so are pots of gold for the information scientists can glean from them. *D. problematicus* was about the size of a northern pike, twenty to thirty inches long, and the 409-million-year-old Canadian fossil preserves a braincase, scales, calcified cartilage, and sharp teeth in the upper and lower jaws. The teeth grew in batteries three to four deep, set up for shedding and replacing—a great indication that "revolver dentition" was part of the original chondrichthyan equipment.

One of the oddest of the odd ancient sharks first appeared in the latter part of the Devonian, the four-foot-long *Stethacanthus*. Males in this genus had what looked like a spike-studded ironing board or mooring bollard on their heads, and long, toothlike scales on the first dorsal fin. In the standard explanation of freakish ornamentation, scientists think the appendage played a role in courtship. Speaking of which, from those earliest days, chondrichthyans practiced internal fertilization. Males developed penis-like structures called claspers that they inserted into females to efficiently deliver sperm. Most of the bony fishes, then as now, practice external fertilization, in which the female expels eggs that the male fertilizes by releasing sperm over them.

In the Devonian planetary configuration, the Laurentia and Baltica landmasses had collided to create Euramerica, while to the south, Gondwana, down near the bottom of the globe, had rotated its western margin clockwise toward the equator. Tetrapods, the first four-limbed vertebrates, were taking tentative steps out of the water, but land plants, having gotten the hang of roots and branches, ruled Devonian terra firma.

The Devonian Age of Fishes ended in a mass die-off, mostly of marine species. Scientists debate what caused it, although they generally agree it was probably a slow-motion event, consisting of a rolling or pulsed series of extinctions that occurred over millions of years. One of the most widely accepted explanations is that the extinctions were due to changes in the atmosphere brought about by the thriving communities of all those respiring vascular plants. Whatever caused the end-Devonian extinctions, *Dunkleosteus*'s formidable armor was no protection, and the placoderms did not survive. But sharks did.

Which brings us to the Carboniferous (359.2–29 mya)—the matchless Golden Age of Sharks. With *Dunkleosteus* out of the way, sharks seized the day as the ocean's apex predator. Not only was their own major predator gone, but also their prey base expanded. Rising sea levels at the beginning of the period favored the development of invertebrates like corals, sea lilies, and mollusks. Some of the ray-finned fishes as well as some sharks evolved crushing teeth to take advantage of these new food sources, while other sharks held on to their sharp teeth, the better to eat all those ray-finned fishes.

The Carboniferous is the only period divided into subperiods, which are the Mississippian (earlier, or lower), and Pennsylvanian (later, or upper). However you divvy it up or whatever you call it, the Carboniferous was crammed with chondrichthyans of all shapes and sizes. A mind-boggling degree of evolutionary experimentation was going on within the group then, or as scientists would say, it was a time of great evolutionary radiation. Sharks were at their peak diversity of all time in the Carboniferous, showing up in all shapes and sizes, from a few inches long to twenty or thirty feet. This is when our cartilaginous friends really started to segregate into the "sharky"-looking chondrichthyans (mostly the elasmobranch side of the class) and chondrichthyans that most spectacularly did not pass the shark "gestalt" test, like the weird and wonderful iniopterygians and petalodonts, which grouped on the holocephalan side of the class, which we will revisit in chapter 12.

Cladoselache, Ctenacanthus, and *Stethacanthus* all squeezed through the Devonian extinctions into the Carboniferous. There they were joined by the likes of *Falcatus*, called the "unicorn shark" for the male's head appendage (for courtship of course), and finally, *Edestus*, which evolved and went extinct in the late Carboniferous. *Edestus* introduced one of the Carboniferous period's boldest evolutionary experiments, the curved symphyseal tooth blade. The term "symphyseal" refers to positioning. Symphyseal structures are located at the symphysis, or point of "fibrocartilaginous fusion," where two bones or cartilages join together. (In biology, joints are also called articulations.) In the case of *Edestus*, and later *Helicoprion*, the symphysis in question is the front junction of the jaw. Symphyseal structures are by their very nature "medial," or centered. A whole new order, the Eugeneodontida, would have to be created for the curved-tooth evolutionary oddballs, although these "eugeneodontids" wouldn't be organized and named until the 1980s, more than a century after the first fossil from this group was discovered.

Forgetting sharks for a moment and arguing from the tetrapodal perspective, the greatest evolutionary innovation of the Carboniferous was the amniotic egg. With this sturdy system, an embryo develops inside a protective membrane or covering that can either be laid in the open air like a chicken egg or retained for the entire gestation inside the mother. Amniotic eggs allowed the four-limbed tetrapods to become fully and securely terrestrial. Maybe those mobile, independent, hungry tetrapods clambering out of the sea to stretch their legs and their potential are what inspired Carboniferous insects to finally evolve flight, and for the first time, the skies, too, began to fill with life.

At the beginning of the Carboniferous, the drifting landmasses of Siberia and Kazakhstania had tagged onto Euramerica. By the end of the Carboniferous, the expanded Euramerica had docked onto Gondwana to form a supersize supercontinent called Pangaea. The embryo-shaped Pangaea extended from the far Northern Hemisphere across the equator to the South Pole and was surrounded by the global Panthalassic Ocean. With Pangaea, the stage is set for the Permian period (299–251 mya), the closing scene of the Paleozoic. *Helicoprion* lived near the middle of the Permian. Some paleontologists say late-early Permian, and others say early-late Permian (scientists love their nuances).

British geologist Roderick Impey Murchison had already named the Silurian and co-named the Devonian periods when he mounted an expedition to the Ural Mountains in 1841, six years before Karpinsky was born. One of the last of the independently wealthy gentlemen geologists, Murchison had only gotten into science after his wife and friends urged him to find something to do other than foxhunting. Privilege positioned Murchison, but his intellect and aggression wrote him into the science history books. He was adamantly and obstinately right, and just as adamantly and arrogantly wrong, on a number of issues. He was right that the Devonian came between the Silurian and the Carboniferous, but he also insisted that life began in the Silurian, instead of a more "primordial" time, as proposed by French paleontologist Joachim Barrande. In the end, Murchison's contributions to geology and higher learning outweighed his stumbles, and he is recognized as one of the nineteenth century's most distinguished earth scientists. To his partial credit, Murchison was among the Geological Society members who urged recognition of William "Strata" Smith, though he would have never suggested membership for the coal-smudged surveyor.

Murchison and two coauthors, French geologist Édouard de Verneuil and Russian Imperial representative Count Alexander von Keyserling, published *The Geology of Russia in Europe and the Ural Mountains* in 1845. It was there, in chapter 8, that they introduced and explained their new term, "Permian":

> *Convincing ourselves in the field, that these strata were so distinguished as to constitute a system, connected with the carboniferous rocks on the one hand and independent of the Trias [Triassic] on the other, we ventured to designate them by a geographical term, derived from the ancient kingdom of Permia, within and around whose precincts the necessary evidence has been obtained. . . . In a word, therefore, our Permian system embraces everything which was deposited between the conclusion of the carboniferous epoch, and the commencement of the Triassic series.*

The "ancient kingdom" Murchison referenced was probably Great Perm, a medieval city settled by the ethnic Komi people.

Great Perm long predated the modern city of Perm, which was built around 1780 as a factory settlement for a large copper operation. The name Perm is thought to have derived from a word in the Uralic language family, *perama*, meaning "faraway land." It was fitting. The province of Perm was so remote that for much of Russian history it was one of the places to which the disfavored were exiled.

Stepping across that bridge of imagination once again, far away and long before the Komi kingdom, there was the Permian realm of Pangaea. . . .

As dawn rose over the vast supercontinent, a ten-foot *Dimetrodon* lifted its solar-collector sail to the morning sun. When it was sufficiently warmed, the confident predator resumed its previous day's hunt along the western shoreline. *Dimetrodon* looked like a dinosaur, with its sail-like crest, clawed toes, bristling teeth, and long, thick tail, but it was a synapsid, a lizard-looking amniote and early mammal relative. Dinosaurs were still million of years into the future. A splash just offshore caught the *Dimetrodon*'s attention as a *Mesosaurus*, a three-foot-long aquatic reptile, took a breath and flipped its crocodile-like tail, diving back under the water to look for its next fishy meal. Out to sea beyond the *Mesosaurus*, the Panthalassic Ocean, the ancestral Pacific, stretched to the horizon and beyond—wrapping around the globe to come back lapping at Pangaea's eastern shore and mingle with the waters of the Tethys Sea, a huge marine embayment in the belly of the supercontinent.

Before the cataclysmic end of the period, it was a good time in the Permian. Life was booming, both on the land and in the sea, where a diversity of species thrived in the warm waters. Currents were simple and slow. Large reefs with corals, sponges, brachiopods, echinoderms, stromatolites, and other reef creatures flourished along the continental shelf. The last of the trilobites were scurrying across the seafloor while the ammonoids and nautiloids and probably unshelled squid jetted around in abundance, eating and being eaten. Marine life had recovered following the end-Carboniferous extinctions, and new species proliferated to fill empty niches. Although the trilobites were on their way out, the bony fish were making themselves at home. Chondrichthyan diversity had taken a shellacking at the end of the Carboniferous, but sharks were still the ocean's top predator. Although cartilaginous fishes—the sharks, skates, rays, and chimaeras—have proven to be tenacious survivors, they would never regain their previous variety. Overall numbers were strong though, and there were plenty of sharks in the sea—including the heir to the *Edestus* throne, the Permian shark king, *Helicoprion*.

While the *Dimetrodon* lumbered south on all fours, a *Helicoprion* cruised north along the coastline, trailing a squadron of squid. The shark was a big beast—not the biggest *Helicoprion* around, but nearly twenty feet long. As big as a great white. As big as a ski boat. It moved through the sun-dappled submarine world with pantherine grace and control, smoothly paddling its tail and maintaining a steady lift using its pectoral fins like wings. Following a route known to wide-ranging sharks like itself, the *Helicoprion* eventually turned east into the narrow Uralian Seaway, probably following cues gathered by the ampullae of Lorenzini clustered, or probably clustered, on its face. Scientists don't know exactly when chondrichthyans evolved these sensory organs—jelly-filled pores holding bundles of extremely sensitive cells and nerve fibers that detect electrical signals—but it's a safe bet they were present in Permian sharks. Modern sharks use their ampullae of Lorenzini to detect prey, communicate with other sharks, and navigate by the earth's magnetic field, so maybe the traveling *Helicoprion* knew that this seaway was a shortcut to the Tethys Sea, a proto-Mediterranean. Our *Helicoprion* made it about halfway down the seaway, which might have been some three thousand miles long, when something happened. It probably didn't die of old age—old age is a luxury usually reserved for domesticated creatures—but die it did, and settle to the bottom. The shark's next major disruption, more than two hundred million years later, would be Russian quarry workers.

Our *Helicoprion*, already lying quietly and well covered by sediment at the bottom of the Uralian Seaway for millions of years, missed the tectonic closing of the seaway and the drama of the Permian extinction. The curtain didn't just come down, the theater collapsed. Some 96 percent of marine species and 70 percent of terrestrial species went extinct in the "Great Dying." Entire taxonomic families perished. Twentieth-century paleontologist Curt Teichert, who comes to play a role in the *Helicoprion* story, wrote this about the Permian extinction: "The way in which many Paleozoic life-forms disappeared towards the end of the Permian Period brings to mind Joseph Haydn's Farewell Symphony where, during the last movement, one musician after the other takes his instrument and leaves the stage until, at the end, none is left." Scientists believe there may have been three distinct extinction pulses with different causes, ranging from gradual climate change to catastrophic volcanic activity. Between the end-Carboniferous extinctions and the end-Permian extinctions, chondrichthyans were seriously depleted. They did survive into the next era of course, but into a class more starkly divided.

What doesn't kill you makes you stronger, and the surviving life-forms staggered back onstage. The show must go on. The Mesozoic era (251–65.5 mya), with its Triassic, Jurassic, and Cretaceous periods, followed the Great Dying. Dinosaurs arose in the Triassic, as did pterosaurs, ichthyosaurs, mammals, turtles, and frogs. By the Jurassic period, Pangaea had broken up into separated landmasses, creating favorable new marine habitats that included expanded continental shelves and shallow-water environments. Jurassic sharks rallied for their second major expansion in diversity ("radiation") around the time the first birds took flight.

In the Cretaceous period, modern sharks, the "neoselachians," radiated rapidly, taking advantage of newly abundant bony fishes. Plentiful and diverse, Cretaceous sharks spread through the earth's newly rearranged oceans as near-shore predators and swift offshore hunters. It's thought that the Cretaceous, which ended sixty-six million years ago, was brought to a close by an asteroid strike that sparked the earth's second-largest mass extinction. All the dinosaurs died out (except for the ones that were on their way to evolving into birds), along with about 50 percent of the rest of earth's species.

Again, the show went on. The Cenozoic era (65.5 mya to the present) began with landmasses in an arrangement of continents that we would largely recognize. The meek mammals inherited the earth from the go-big-or-go-home dinosaurs. Feeling an old pull, mammals put out to sea and grew into large flipper-footed animals like sea lions, which allowed the sharks who ate them to grow even larger. Megalodon, which was about the size of a railroad boxcar, was a Cenozoic shark.

The Cenozoic has only two periods, the Paleogene and the Neogene. Those have been divided into seven epochs, which might sound more familiar. From oldest to youngest the Cenozoic epochs are the Paleocene, Eocene, Oligocene, Miocene, Pliocene, Pleistocene, and Holocene. We are living in the Holocene, although some scientists convincingly and alarmingly argue that we are actually in a new "cene," the Anthropocene, with plastic, aluminum cans, nuclear fallout, and even mass extinctions, as the signatures of our age, to be found in the rock layers and puzzled over a million years in the future, if there is anyone left to puzzle.

Red Bull cans would likely have been as puzzling to Alexander Karpinsky as the fossil tooth whorl he unwrapped on that fine spring day in 1898. Maybe he smiled at this most pleasing gift from Ural-Batyr, by way of Bessonov. He had no idea what it was.

6

KARPINSKY MAKES THE CALL: *HELICOPRION*

It is remarkable how many of the dead shark workers are, and have always been, scientists of powerful intellect and remarkable originality. This is all the more remarkable because the material is horribly scrappy and rare.
—Augustus (Toby) White, writing on palaeos.com

ALEXANDER KARPINSKY SCRATCHED HIS HEAD, OR AT LEAST TUGGED HIS BUSHY, graying goatee. It's easy to imagine him circling his work table scrutinizing Bessonov's fossils, bending in to note a detail and leaning out to consider the larger whole. We know Karpinsky relished "problematic" fossils like the ones before him, no doubt for the academic challenge and geological advances to be gained through each new layer of understanding. Maybe he also loved their mystery and allure. Fossils aren't just rocks, they're enigmatic envoys with coded intelligence from deep behind the lines of time—and Karpinsky was one of the best code breakers of the day. Even so, he was stumped. He had never seen anything like these fossils in all his days of fieldwork, research, and steady knocks on the door from associates bearing inscrutable specimens asking, "Professor, do you have a minute?"

Maybe he was alone in his lab on that spring day in 1898, happy to quietly contemplate these most surprising fossils: one broken whorl and one intact whorl, both nine to ten inches in diameter, plus three fragments. The broken whorl was missing a large chunk, giving it the shape of a Roman helmet. Even with its missing section it showed three and a half volutions (full turns) and had 146 teeth, either preserved or clearly indicated. The second whorl . . . simply astonishing. Unbelievable. Beautiful. The second whorl was spectacularly intact, with just over three full volutions and 136 teeth, many with the crowns exquisitely preserved. Karpinsky could have put his finger in the center of that whorl and traced the spiral all the way out, with his finger bumping over the tightly packed teeth, never missing a beat. The smallest teeth were at the center—the tip of his finger would have covered several—and grew progressively larger, up to about an inch and a half tall at the perimeter.

Karpinsky's mind would have been churning, trying to lace some of his large store of existing knowledge into this open latticework of unknowns. He was confident he knew how old the fossils were. He had studied the strata in that part of the Urals for years, focusing on the abundant fossil ammonoids in the region's outcrops and quarries. If luck would have had it differently, he might have been the one to discover *Helicoprion* himself, considering all the time he spent prying fossils from rocks in the area. Karpinsky knew the specimens lying on his table were from the Artinskian age of the Permian period. Karpinsky had identified and named the Artinskian age himself twenty-four years earlier, for the small city of Artinsk (now Arti), forty miles southeast of Krasnoufimsk.

Karpinsky also knew the spiraled geometry quite well, from his long-standing interest in the logarithmic spiral's occurrence in nature, particularly in ammonoid shells. As it grows, a logarithmic spiral retains its shape while expanding its volume. Unlike the Archimedean spiral of screws and watch springs, in which each turn measures exactly the same as the one before, the logarithmic spiral opens slightly with each turn in a geometric progression. In such mollusks as the nautiloids and ammonoids, the logarithmic expansion of the shell allows the animal to grow without changing shape. He believed the spiral growth pattern must be an essential economy of matter and energy, since the form occurred in such a wide range of organisms, including the shells of giant ammonites and single-celled foraminifers, those fascinating, paleontologically important subjects of Kingdom Protista, whose shelly backs built many a limestone reef.

Karpinsky didn't think for a second that the spiraled structures on his table were forams, or ammonites, or any kind of shells at all. These were something else. Something new. Something vertebrate. He was sure of it.

Leaving Karpinsky to think his scientific thoughts, we can slip off the shackles of scholarly propriety and look over his shoulder for the pure pleasure of absorbing the lines, curves, and textures of the fossils on the table. The intact whorl was a graceful, pleasing mathematical spiral planted with an orderly battalion of teeth neatly tucked together. For a fossil, the whorl possessed a remarkably clean symmetry and gear-like precision. The tooth crowns looked like serrated knife tips that could tear any flesh put before them, right there, right then. Perhaps Karpinsky, who was known to be quite musical, hummed as he studied the fossils. He was a methodical and careful scientist who considered prevailing scholarship, hung his own ideas inside that frame, and crafted a hypothesis that didn't contradict known facts, on which he had an assuredly firm grasp. These extraordinary fossils, however, were completely new to him. His goal with the odd, spiraled specimens was to identify the most probable and scientifically grounded conclusion, and that was the scrupulous route he took in his comprehensive investigation of the fossils from Krasnoufimsk.

The monograph Karpinsky published on *Helicoprion* in 1899 (in German, the favored scholarly language of the day) was more than one hundred pages long, with four plates and seventy-two figures including photographs, line drawings, and microscope-thin sections. The groundbreaking opus, titled, "About the Remains of Edestiden, and the new Genus *Helicoprion*," began with a short, three-paragraph introduction. First, Karpinsky credited the school inspector from Krasnoufimsk, Mr. A. Bessonov, with sending him the fossils in the spring of 1898. In the second paragraph Karpinsky asserted that, despite their ammonite-like spiraling, the whorls belonged to a vertebrate animal, specifically that most preeminent group of vertebrate animals, the sharks. He respectfully submitted in the third paragraph that the new fossils represented a previously unknown genus, which he was naming *Helicoprion*—from the Greek *helico,* "spiral," and *prion,* "saw."

There would be plenty to argue about over the next hundred-plus years, but no one would dispute the fact that these fossils were indeed a brand-new genus of ancient shark. So in 1899, *Helicoprion* was enshrined on the short, sanctified list of named, extinct creatures.

Following his brief introduction, Karpinsky presented a seventeen-page review of the *Edestus* genus. It made sense for Karpinsky to start there, since that unprecedented beast had been the starting point in all this curve-toothed shark business. (Tooth, fin spine, whatever.) *Edestus* was one of the knowns for Karpinsky's frame. He covered the handful of species that had been identified since Leidy's original description of *E. vorax*, and discussed what scientists knew, and what they speculated, about the strange fossil structures. Of special interest to Karpinsky was the Australian fossil, *E. davisii*. Habitually tactful and generous when disagreeing with other scientists, Karpinsky acknowledged both Henry Woodward's in-depth analysis of the Australian fossil fragment, and his reasonable decision to assign it to the *Edestus* genus. The esteemed British expert just hadn't had enough information, Karpinsky offered—but now here it was. Mr. Davis's fossil was far more similar in structure and form to *Helicoprion* than to *Edestus,* even in its broken state. Therefore, Karpinsky provisionally transferred *davisii* into his new genus (a legitimate move in the chess game of taxonomy). Enough distinctions remained between the Australian and Uralian fossils that he created a new species name for the Russian specimens, *bessonowi*. Where there had been none, now there were two: *Helicoprion davisii* and *Helicoprion bessonowi*.

In the rest of his hundred pages, Karpinsky left no stone unturned. He studied the rock in which the fossil was found, put the rock and the fossil into the context of known paleogeographical data, searched for comparisons among extinct and extant sharks, scrutinized the form and structure of the whorls, and otherwise noted every last detail. His inclusion of thin sections in the monograph was impressive. Advances in polarized light microscopy were quite recent, and few vertebrate paleontologists were using this technical innovation in their research. In a thin section, a sliver is cut from a rock or fossil so it can be viewed under a microscope with polarizing light filters that make individual minerals pop out in distinct patterns and colors for ready identification. (Besides being useful, thin sections can be very pretty.) Karpinsky's thin sections of the fossil's toothy parts revealed vasodentin covered by vitrodentin. Dentin, you'll recall, is the hard, calcified tissue that makes up the bulk of a tooth. Vasodentin is a type of capillary-fed dentin common in the teeth of non-mammalian vertebrates (like fish), and vitrodentin is an enamel-like substance that covers the tooth crown in nonmammalian vertebrates. Both of those substances are also found in chondrichthyan dermal denticles, those elegant little pseudo-teeth at the heart of the *Edestus* fin spine argument. Regardless, Karpinsky felt the evidence most strongly showed that *Helicoprion*'s whorl was a "dental apparatus." He suggested that the inner teeth were small because they were *Helicoprion*'s first teeth, its baby teeth, and that teeth got bigger as the animal added more during its lifetime. Instead of shedding teeth like other sharks, Karpinsky believed *Helicoprion* evolved a mechanism to retain its teeth by reeling them into an oh-so-biologically-economical spiral. Older, smaller teeth tucked up inside; new, bigger teeth on the outside to keep up with the growing body.

Throughout the paper, Karpinsky admitted what he didn't know, with the confidence of a man at the forefront of his field. In a sense of collegial unity and shared purpose with the scientific community, Karpinsky not only fully described what his research had revealed about *Helicoprion*, he also outlined and discussed his questions, doubts, and uncertainties. He wrote that he had highlighted especially dubious interpretations in order to encourage future researchers to push *Helicoprion* studies in those directions, and said he did his best to express his ideas, which had shifted over the course of his work.

With his evidence-based observations thoroughly recorded, Karpinsky ventured into theoretical territory. Without theories, evidence just lies around like a heap of disarticulated bones that will never leap up to tell their life stories. Theories—ideas, hypotheses, conjectures—animate evidence so it can be viewed from all sides, in all plausible possibilities. Well-reasoned speculative ideas are intellectual exercises—scientists putting on their pith helmets to go exploring. So off went Karpinsky. He did believe that the whorl was a mass of teeth. But he shared the same reservations Newberry had expressed trying to imagine the enormous *E. giganteus* fossil inside a shark's mouth. How could that possibly be? How would it actually work?

At the end of Karpinsky's monograph, following page after page of painstaking scientific description, the elephant in the room was still the question of where that spiked spiral of predatory menace fit on this most mysterious creature. Nodding to Newberry's fin spine proposition that the *Edestus* structure must have been imbedded in soft tissue, Karpinsky theorized the same for *Helicoprion*, although he tried to split the difference by suggesting that the whorl originated in the upper jaw, then pushed out of the mouth as it grew bigger. Consumed as he had been in his study of this strange beast, Karpinsky emerged at a new crossroads of knowns and unknowns, standing alone with a shimmering shark mirage. He squinted hard and intrepidly climbed out on a limb to draw what he saw: a self-possessed shark with a bristling buzz saw curled alarmingly over its snout as a powerful weapon. You have to start somewhere, and Karpinsky definitely got it started.

———————

The spirited, sometimes cantankerous debate that followed the release of Karpinsky's 1899 *Helicoprion* monograph was, as Russian geologist and Karpinsky biographer Sergei I. Romanovsky put it, one of the most colorful pages in the history of science. Arguments, ideas,

and moments of scientific sniffiness flew back and forth across
oceans, creating a de facto forum of international scientists who
wrangled over the enigmatic fossils for decades. Something about
Helicoprion—then as now—ignites an intellectual and imaginative
compulsion to jump in and figure it out. The whorl-toothed shark
is the tight-lidded jar everyone grabs out of everyone else's hands.
I can get this! No, here, give it to me! Everybody wants a crack at it,
everybody has an idea.

An amateur scientist was the first to grab the *Helicoprion* jar,
despite the fact that there at the cusp of the nineteenth and twentieth centuries, scientific research was almost universally conducted
by paid professionals. Ernest van den Broeck, a Belgian financial
agent whose interests included paleontology, geology, religion, Jap-
anese culture, and garden design, published a sketch in the bulletin
of the Belgian Society of Geology, Paleontology, and Hydrology
showing the whorl curled back into the lower jaw. A good geologist despite his amateur status, van den Broeck was attracted to
the earth sciences for their speculative arguments and contributions to public debate, particularly in the areas of evolution and the
"antiquity of man." Van den Broeck's guess was as good as anybody
else's over the next dozen years. At least he put the whorl in the
mouth. But like the rest of the paleontological workers in that de
facto *Helicoprion* forum, for every detail they got right, they got that
many more wrong. In van den Broeck's case, he likened the whorl
to a radula, the minutely toothed feeding structure found in many
mollusks. Sea slugs use their radula to scrape algae off rocks, while
predatory cephalopods use it in conjunction with their keratinous
beaks to shred meat. Van den Broeck's sketch bore a striking resemblance to biological diagrams showing the radula in a snail. Also

appearing in that Belgian bulletin was a sketch by Belgian geologist G. Simoens, showing the whorl pin curled at the end of what looked like a thresher shark's tail.

The first American to jump into the *Helicoprion* pool was paleontologist Charles Rochester Eastman, a professor of geology and paleontology at Harvard and Radcliffe. "Few will be prepared to admit," Eastman wrote of Karpinsky's intrepid reconstruction, "that this highly fanciful sketch can be taken seriously, and, therefore, the least said about it the better."

Eastman may have been feeling cranky about more than Karpinsky's saw-bladed shark. In the same month his criticism was published in *The American Naturalist*, Eastman shot and killed his brother-in-law during some Fourth of July target shooting. While colleagues were reading his *Helicoprion* review, Eastman was awaiting possible indictment for murder. Although Eastman clearly disdained Karpinsky's reconstruction, he wasn't universally dismissive. "It is a significant and decidedly unwelcome truth," he wrote, "that not one in a hundred essays on paleontological subjects receives anything like the elaborate care and finish which Dr. Karpinsky, the Director of the Imperial Russian Geological Survey, has bestowed upon the remarkable [fossils] which he describes under the name of '*Helicoprion.*'"

Eastman agreed with Karpinsky that *Helicoprion* was a shark closely related to *Edestus*, but agreement ended there for the Harvard man. Ever since *Edestus vorax* had been described in 1855, the most vigorous voices in the paleontological community had argued that the *Edestus* fossils were fin spines, regardless of Joseph Leidy's original instinct that they were teeth. Now, Karpinsky's inclination to call *Helicoprion*'s whorl a tooth structure reinvigorated that whole debate. Eastman prominently cited the opinions of his associate Bashford Dean, a vociferous, confident, and some said brash advocate of the fin spine argument. Dean himself had named a new *Edestus* species just a few years earlier, *E. lecontei*, based on a fossil found in Nevada. Dean's idiosyncratic contributions to science included pioneering work that combined embryology and paleontology to study the evolution of chimaeras (the ratfish), which he believed to be "highly specialized offshoots from some lowly branch of the true sharks." He was also an expert on armor, establishing the Metropolitan Museum of Art's Department of Arms and Armor. Dean wasn't afraid of a fight, and had continued to be unflinching in his assertions that the *Edestus* structures were fin spines.

Dean was convincing, Eastman agreed, but at the same time, what about those other paleontologists who had been impressed by the similarity between the *Edestus* structures and *C. megalodon* teeth? It seems as though Eastman wanted to believe the fossils were teeth, but a shark with such unwieldy whorls in its mouth would be a monster. In the end, he balanced on the spine-versus-tooth fence. "The question, then, as to whether these objects are segmented spines or teeth would seem to remain as puzzling as ever, in spite of the abundance of new light thrown on their structure by Karpinsky's studies."

———

British paleontologist Arthur Smith "A. S." Woodward offered a more concordant response to Karpinsky's paper in *Geological Magazine*, edited by Henry Woodward himself. In addition to editing the magazine, Henry Woodward was also the keeper of geology (department head) at the British Museum of Natural History, where

A. S. Woodward had achieved the position of assistant keeper. The men were not related, and when A. S. began working at the museum seventeen years earlier, just after his eighteenth birthday, Henry and Arthur agreed that to avoid confusion, Arthur would always include "Smith" (his mother's maiden name) in identifying himself. A. S. had recently completed the monumental three-volume *Catalogue of the Fossil Fishes in the British Museum,* a project that established him as one of the world's foremost experts on fossil fish.

"Paleontologists are indebted to the eminent Director of the Imperial Russian Geological Survey for one of the most exhaustive memoirs on a fossil ever published," A. S. wrote in the 1900 review. He praised the piece as "a model of what such a work should be—thorough from every point of view, geological, chemical, and biological."

A.S. agreed that Karpinsky "rightly judged" *Helicoprion* to be a previously unknown genus of ancient shark. "*Helicoprion*, to a superficial observer, looks much like an ammonite," he wrote, "but on closer inspection it is easily recognizable as a spiral consisting of teeth firmly fixed together by their bases." He had no trouble using the word "teeth," although he did acknowledge the controversy. "The rival theories by which *Edestus* has been sometimes ascribed to the jaws, sometimes to the external dermal armour of a shark or skate, can thus be discussed again in the light of important new facts."

Some of Woodward's important new facts came from Karpinsky's monograph, while others derived from recent discoveries by eminent Scottish paleontologist R. H. Traquair that seemed to support the existence of even older Paleozoic sharks with curved, symphyseal tooth plates—symphyseal, remember, means positioned on the center line, aligned with the symphysis, or junction, at the front of the jaw where the two sides come together. Woodward used the light from those new facts to push adventurously on. What if the whorl *had* been inside the mouth, instead of curled up over the snout like Karpinsky pictured it? Why not? What if the sharks possessed not just one but multiple whorls? Woodward mused that it would be interesting to know if Karpinsky's whorls and fragments had been found close together, or maybe even in association, as such details might offer a clue. Unfortunately, that information was lost to a quarry worker at Divya Gora. But what if? After all, modern sharks had multiple rows of teeth. Why not multiple whorls?

Looking for possible analogs to support his idea of multiple whorls in the mouth, Woodward turned to the frilled shark (*Chlamydoselachus*), a living but very ancient creature with rasp-like rows of needle-sharp, three-pronged teeth. Frilled shark teeth look like squid lures that anglers use to jig for squid—sharp, crowded barbs perfect for snagging flesh. Good plan, since frilled sharks eat a lot of squid. Woodward pointed out that some of *Chlamydoselachus*'s tooth rows exhibited bilateral symmetry, and the frilled shark's mouth was full of them. As he wrapped up his review of Karpinsky's research, the core question A. S. Woodward posed to the paleontological community was, which was weirder? A single whorl of external teeth curled up over the snout—or the whorl, maybe multiple whorls, inside the mouth?

"The conception of a gigantic shark armed in both jaws with several series of teeth like those now described under the name of *Helicoprion* is, indeed sufficiently startling," he concluded, "but it seems to us more likely to be realized than the hypothesis which Dr. Karpinsky's most interesting researches have led him to propose."

Interesting enough researches, in fact, to escape the academic journals and surface in the July 1900 issue of *Popular Science*. The widely circulated American magazine could be counted on to cover the relatively new and hot topic of evolutionary science, along with its reporting on inventions like the telephone, phonograph, and automobile. The *Helicoprion* article featured a photograph of the fossil, as well as Karpinsky's snout whorl reconstruction and a line drawing of the *Edestus giganteus* fossil. The writer nicely summarized existing knowledge of *Edestus* and *Helicoprion* but couldn't share with readers the most curious detail of exactly where the whorl fit. That could only be conjectured. It might possibly have been curled in the tail, "this part being armed with prickles" in some living sharks, skates, and rays. "More probably, however, judging from the character of the teeth, it was a prolongation of the snout." Modern sharks do shed their teeth, the writer acknowledged, but maybe in *Helicoprion*'s case, instead of dropping out of the jaw, teeth were carried forward in front of it. "Hence, the convoluted muzzle could be driven directly ahead, the strong teeth making it a most formidable assailant." *Popular Science* seemed to be backing Karpinsky's vision, and popular opinion was lining up behind teeth, wherever they might fit.

Of course the beauty of science lies in the fact that it isn't a popularity contest. The curve-toothed shark questions were far from settled, and the lid of unknowns remained tightly screwed on the *Helicoprion* jar. It was still a story of blind men and the elephant. The blind men needed more clues. They needed more sharks.

7

A SHIVER OF SHARKS

*That's just how science works: it's not about who's right or wrong but what
the truth is, and how we can best determine the value of new discoveries to
understanding the bigger picture of . . . evolution.*
—John Long, from *Dawn of the Deed*, 2012

CHARLES ROCHESTER EASTMAN RESUMED HIS THREAD OF THE CURVE-TOOTHED SHARK
conversation in 1902. He had finally been acquitted on all charges in
the murder of his brother-in-law, but not before serving eight months
in jail and enduring a sensational, very public, three-week trial. After
the verdict was read he declined to make a statement to the press, but
greeted his wife and family, shook hands with each juror, and went
to the jailhouse to thank the sheriff. Then C. R. Eastman got back to
work.

His thinking had apparently clarified, because Eastman returned to
the scene firmly in the *Helicoprion* tooth camp and running at the front
of the pack in the race to find new evidence, which meant new specimens. With the buzz in the scientific press, curators had started spot-
ting curve-toothed fossils in long-shut museum drawers. It's like when

you've never seen a thing before—say a morel mushroom—then someone points one out under a tree. Suddenly you start seeing them yourself, because now you know what you're looking for. You know there *is* something to look for.

Eastman received one such long-overlooked specimen from his friend J. S. Kingsley. The unidentified fossil belonged to the Tufts College Museum in Boston, and all Kingsley knew about it was that it had been acquired many years before by a since-deceased professor. The fragment, which was more curved than *Edestus*, was an impressive nine and a half inches long and appeared to have once held twenty teeth. Most were broken off but four remained fairly well preserved, with the largest a great-white-shark-worthy three inches tall. Eastman described the fossil in *Geological Magazine*, erecting a new genus for it as an intermediate position between the slightly curved *Edestus* and the fully spiraled *Helicoprion*. He named the new genus *Campyloprion*: *campylo* meaning "curved," and *prion*, "saw." Curve-toothed saw. Eastman proposed that Mr. Davis's fossil most rightly belonged in his new *Campyloprion* genus, and while he was at it, he rustled Bashford Dean's *Edestus lecontei* into the *Campyloprion* corral. That made the third proposed assignment so far for *davisii*: *Edestus davisii* to *Helicoprion davisii* to *Campyloprion davisii*. The 1902 article was short—Eastman's primary goal was to stake his naming claim on *Campyloprion* and bring the already-known Paleozoic elasmobranch *Campodus* into the curve-toothed shark fold. But as promised, he followed up quickly with a more detailed article, published that same year in the *Bulletin of the Museum of Comparative Zoology at Harvard College*. One of the key observations Eastman brought to the fore in that paper articulated an important distinction and key difference between *Edestus* and *Helicoprion*: the roots of *Helicoprion*'s teeth formed a unified structure—in other words, the teeth shared one continuous root—while the roots of *Edestus* teeth remained clearly segmented. He was absolutely right. He further noted that the tooth bases of *Helicoprion*, *Campyloprion*, and *Campodus* bent in one direction from the crowns, whereas the *Edestus* tooth bases bent the opposite way. This observation would prove pivotal to the tooth-versus-fin spine debate.

Eastman stood firm on the side of teeth, and dispatched previous theories and suggestions from Newberry, Richard Owen, Bashford Dean, and others who had argued that the *Edestus* structures were fin spines. "The curved or coiled 'spines' of *Edestus* and *Helicoprion*," Eastman wrote, "are not dermal defenses at all, but veritable teeth." With that question out of the way, at least in his own view, he admitted that the nature, function, and relations of the tooth structure remained "highly problematical." Which is not to say he didn't have ideas—he did, many springing from the *Campodus* fossil he had received from Professor Edwin H. Barbour of the University of Nebraska at Lincoln. (Eastman calls this particular fossil a *Campodus* here, but the specimen would eventually be moved out of the *Campodus* taxon.) One of the notable things about the specimen—a lovely, highly curved fan of eleven teeth prepared with "skill and zeal" by Barbour's sister—was that it had been found in association with more than four hundred tiny, flattish "pavement" teeth that could have served as a grinding surface. With such pavement teeth to work in concert with the whorl, and despite A. S. Woodward's frilled sharks, Eastman surmised there was only one whorl per animal, with a "chiefly masticatory" function. Eastman gave a little dig, and a little nod, to Fanny Hitchcock as he came to his conclusions. Her "interpretation was at fault," he wrote, when she compared *Edestus* and *Onychodus*. On the other hand, "her reference to the median line, in front of the lower jaw, was a close approximation to the truth, as has been finally revealed through a study of *Campodus*." With some apparent relief, Eastman declared that he and his fellow paleontologists could finally "avoid the rather formidable conception of giant sharks in the Carboniferous, armed each with a mouthful" of whorled tooth blades.

Eastman acknowledged that scientists knew these sharks only by one or a few whorls each, most of which were poorly preserved fragments. Nevertheless, "these forms taken together constitute a remarkable series, in which the progress of evolution is readily traceable." It could have all started in those older chondrichthyans with the knob-like symphyseal tooth plates meant for crushing—

> *As these teeth became enlarged through specialization in various genera, the difficulty of accommodating them without their proving an encumbrance to the creature was overcome by the simple device of coiling—the same mechanical contrivance which had already been carried to a remarkable perfection amongst the Nautiloids, and was never afterwards abandoned amongst Ammonites except with disastrous or fatal results. In this parallelism between the coils of* Helicoprion *and involute [coiled]* Cephalopods *we observe the culmination of efforts expanded along a certain direction, the design being to accommodate a large number of segments in a minimum of space and at the same time to provide for maximum rigidity.*

With this sort of analysis and his dedication to the topic, Eastman established himself as a leading voice in the de facto curvetoothed shark forum. The conversation had gotten so lively that in

1905, Eastman was compelled to round it up in *The American Naturalist* with an index of all the papers so far published on the topic. The list included forty-four papers written by twenty-four authors from the United States, Russia, Europe, and Japan. "Hardly has some form of animal life, previously unheard of and apparently unique, been brought to light," he wrote, "when identical or closely related types are reported from remote regions." The morel mushroom effect.

According to Eastman, the majority of writers agreed that *Helicoprion* and *Edestus* fossils were peculiarly modified teeth that became fused into a curve or spiral. Dissident voices persisted, however. "The most recent communication that has appeared on this subject strikes a slightly discordant note," Eastman reported. Discordant, at least, to his own point of view. British paleontologist Edwin T. Newton had suggested that paleontologists might be wrong to link *Helicoprion* and *Edestus* so closely together. Newton could believe that *Helicoprion*'s whorl was a spiral of teeth, but he could not see the same for *Edestus*. By the looks of it, Newton argued, the *Edestus* structure must have been a defensive dorsal fin spine. Eastman's aggravation with Newton's stance seeped out between the lines of his response, in which he pointed to an "almost perfect" correspondence between *Edestus* and the legitimate whorl that he was illegitimately calling *Campodus*. He seemed to fairly shout up from the page, *Put them side by side and look harder, man!*

While Eastman and the rest argued about taxonomic relations and took their endless measurements of terminal apices and apical margins, the masses just wanted to see the monster shark. Public interest in prehistoric creatures had continued to grow, from the days of Richard Owen's Crystal Palace dinosaur models through news coverage of Cope and Marsh's Bone Wars, an abundance of new fossils, improving museum exhibits, and popular books like *Tiere der Urwelt* (Animals of the Primeval World) by German author Wilhelm Bölsche. The first artistically styled, full-body rendering of *Helicoprion* appears to have been produced for *Tiere der Urwelt*, by a German artist we only know as "F. John." Bölsche was primarily a poet and novelist, but he was also a great popularizer of science through his lavishly illustrated books and articles on the natural world. He wasn't a scientist, and some argued he wasn't even much of a naturalist, but he seemed to think the whorl might have been displayed like an ammonoid decoy to lure in unsuspecting cephalopods. A cephalopod would have to be terribly unsuspecting indeed, since the "decoy," as F. John depicted it, looked like it was caught in the shark's mouth. Regardless of logic or lack of evidence, the illustration showed the whorl knurled like an ammonoid shell rather than toothed, and curled under the shark's chin instead of spiraling over its snout, as Karpinsky had it. The eel-shaped body resembled a frilled shark. Even before the book came out, the *Tiere der Urwelt* illustrations were printed by the Reichardt Kakao company in a series of collector cards, probably given away in boxes of cocoa. Correctly rendered or not, *Helicoprion* claimed its place around the world's breakfast tables along with the novelty cards depicting dinosaurs, plesiosaurs, ancient rhinos, and the far less primeval but no less extinct dodo.

In 1907, Indiana-born paleontologist Oliver Perry Hay described and named another new "prion" genus, from a fossil fragment found in Bear Lake County, Idaho. Hay named it *Lissoprion*, but the specimen would be correctly identified a few years later as

Helicoprion—the first American *Helicoprion,* our whorl-toothed shark of Idaho.

A Canadian geologist and mining engineer named W. F. Ferrier sent the fossil to Hay, a fragment of three nested teeth, the largest of which was nearly two inches tall. The fossil wasn't intact enough to tell the degree of overall curvature, but Hay thought it more closely resembled *Helicoprion* than *Edestus* because of the way the tooth crowns were fused. He was more correct than he realized on that detail. Hay named the specimen *Lissoprion ferrieri.*

Ferrier, who had served for years as an assistant to Sir William Dawson of the Geological Survey of Canada, was in the United States working as an engineer in the new phosphate mining industry. Fortunately, phosphorus wasn't all Ferrier was interested in. It's true that first and foremost he was a minerals man, a dogged collector known to walk into mine offices and ask for specimens. The minerals he amassed over his career helped build collections at the Royal Ontario Museum, the Smithsonian Insitution, and many others. Even so, Ferrier's mineralogical focus didn't blind him to other treasures, and the enterprising Canadian deserves credit for his intentional and conscientious collecting. He didn't just pick up the biggest and best fossils to catch his eye, but also gathered as much related material as he could, including invertebrate fossils from the same deposits.

As Ferrier's fossils foreshadowed, Idaho's phosphate fields would yield more *Helicoprion* fossils than any other site in the world. To understand why, let's rewind to the Permian period, to that *Dimetrodon* ambling along Pangaea's western shore while Karpinsky's *Helicoprion* swam north toward the Uralian Seaway. Pangaea's long coastline, which stretched nearly from pole to pole, was indented with bays, inlets, and interior seaways—big marine embayments akin in size and shape to

Puget Sound or the Gulf of California. The basin most pertinent to this new American chapter in our *Helicoprion* story was the Phosphoria Sea. Just as *helicoprions* were drawn to travel down the Uralian Seaway, something drew them into the Phosphoria Sea, situated just north of the equator. It may have been a breeding or nursery area, or it may have simply been a good place to find food. (This wasn't the more famous Western Interior Seaway, or Inland Sea, of the Cretaceous period that bisected North America like a great plesiosaur-infested moat. That was still some 170 million years in the future, long after the supercontinent Pangea had broken apart.)

From its core in what is now southeastern Idaho, the Phosphoria Sea extended some three hundred miles west and at least four hundred miles southwest, covering an area the size of four Lake Superiors. While brachiopods, clams, and reefs with corals and sponges flourished in other marine environments during the Permian, they weren't a major part of the Phosphoria ecosystem. During much of its existence, the Phosphoria had a bottom of oxygen-starved (anoxic) muck that if scooped up in a bucket, would likely have reeked of decomposition, with hints of sulfur and methane. Gloppy muck notwithstanding, the Phosphoria was an extremely productive and welcoming environment for free-swimming animals that lived above the ooze. It would have been a nicely green sea full of fish of all sizes, including a healthy population of whorl-toothed sharks and cephalopods of all descriptions, such as ammonites and coiled and straight-shelled nautiloids. (Don't look for *Edestus,* remember, since that group didn't survive into the Permian.) Despite the presence of those complex and interesting animals chasing each other around, the primary author of the Phosphoria's noteworthy geological story was phytoplankton.

It's not a dramatic tale, but epic in its own way. Most phytoplankton are microscopic, photosynthetic marine plants that live in the upper sunlit part of the water column. They supplement the energy they get from photosynthesis by absorbing inorganic nutrients like phosphates, nitrates, and sulfur, which they convert into proteins, fats, and carbohydrates. Phytoplankton are conversion machines, transforming minerals into food for themselves and in turn for other organisms the way cows turn grass into protein for wolves and humans. In a balanced ecosystem, phytoplankton provide a direct food source for everything from baleen whales (which didn't exist in the Permian) to jellyfish (which did), and they also support the food chain by feeding small fish, which in turn feed bigger fish. Phytoplankton live only a few days on average, so they're constantly raining down onto the seafloor where their little dead bodies accumulate as sediment and release nitrogen and phosphorus back to the environment. (Dead fish and other animals also release minerals, but in the Phosphoria, phytoplankton out-contributed the higher life-forms perhaps as much as one hundred to one.)

A seafloor is usually busy with scavenging creatures that recycle this deluge of detritus. Conditions aren't always favorable to benthic animals (bottom dwellers) though, like when there's not enough oxygen for them to survive in the substrate—as was the case throughout much of the Phosphoria's history. What was bad for bottom-dwellers was a boon for phosphate miners and paleontologists. The relatively undisturbed substrate and steady accumulation of organic materials allowed phosphates to form as a sedimentary deposit—a rare occurrence that makes the Phosphoria one of the most extensive known deposits of phosphate rock in the world. And the anoxic environment reduced decay, improving the odds of fossilization in the larger animals that died and drifted down to settle in the muck.

Phosphates are primarily mined for use as fertilizer and in weed killers but are also used to supplement animal feed, and as an additive in biscuits, processed meats, toothpaste, detergents, carbonated sodas, ceramics, and cosmetics. Who knows how many *helicoprions*, brachiopods, ammonoids, crinoids, and other mineralized creatures have been crushed along with the copious tons of phosphate ore trundled off to nourish those famous Idaho potatoes, which might have comprised your last order of french fries? Of course mining is also the reason we have as many *Helicoprion* fossils as we do. Or most specifically, observant and interested mining engineers and mine workers, beginning with W. F. Ferrier and the first three-toothed *Helicoprion* fragment he sent to Oliver Perry Hay.

Ferrier kept looking, and sent Hay three more shipments of fossils from Idaho and Wyoming. In one of the crates was a relic worthy of Ural-Batyr—the world's second intact *Helicoprion* whorl (third, if you count Karpinsky's Roman helmet specimen). No matter that Hay was still calling it *Lissoprion*, we know what it was. In a 1910 paper, Hay described the very deteriorated but beautiful spiral—six inches in diameter with two and a half volutions and an estimated eighty-six teeth. In that same paper, Hay declared that Bashford Dean's *Edestus lecontei*, which Eastman had recently snagged for his *Campyloprion* genus, was really yet another new prion genus. Since Hay saw, or at least recognized it first, he got to name it: *Toxoprion*, *toxo*, meaning "bow" (like an archer's bow), and *prion*, "saw." Bow-shaped saw. Unlike *Lissoprion*, which only survived as a synonym for *Helicoprion*, Hay's *Toxoprion* would withstand the scrutiny of time to persist as a valid genus.

Hay compared and contrasted the *Edestus*, *Toxoprion*, *Lissoprion*, and *Helicoprion* genera, mostly in the context of their teeth. Roots of teeth, blades of teeth, teeth greatly developed; teeth high, cutting edges smooth. Then without warm-up or warning, he started to poke holes in those teeth. If they really were teeth, why didn't they show obvious wear? If they didn't show wear, how could they have been in the mouth? And what about all those ray species with spines? The ribbontails and sting-arees and whiptails? On page fifteen of his seventeen-page paper full of teeth, Hay sprinted back over to the fin spine camp like a man suddenly too thirsty to think anymore. "*Edestus, Lissoprion, Helicoprion*, etc., may for the present," he punted, "be most easily disposed of by supposing that some ancient elasmobranchs developed . . . not a single [dorsal] spine, but a succession of them." Might as well make it a double.

Ever the scientific good sport, Karpinsky illustrated Hay's ideas in a 1911 paper, showing a knightly looking *Helicoprion* with one whorl furled above its back not too far behind its head, and another behind that, sprouting out the back. To more lasting effect, Karpinsky con-tended in that paper that the specimens Hay had been calling *Lisso-prion* in fact belonged to the *Helicoprion* genus. Welcome to America, officially, *Helicoprion ferrieri.*

————————

The very next year, a spectacular *Edestus* fossil described by Hay him-self finally laid the fin-spine-versus-tooth debate to rest for good (or mostly good). R. A. Peterson, an amateur collector from Iowa, sent the fossil to the recently opened Smithsonian National Museum of Natural History, where it was handed off to Hay, who was helping describe the museum's vertebrate paleontology collection. Peterson said a coal miner from Lehigh, Iowa, found the fossil eighteen years earlier in a layer of black shale 165 feet below the surface.

The arresting fossil appeared to capture two curved *Edestus* tooth blades in mid chop. Even in its broken state, the larger blade was over seven inches long, with six strongly serrated teeth over an inch tall. The slightly less-curved second blade was about six inches long with a similar number and size of teeth. This was a very big deal. Not only did the fossil capture a pair of blades together, it preserved a portion of the cranium, offering clear evidence that the blades had been associated with the shark's head. They were not dorsal fin spines. Hay named the new *Edestus* species *mirus*—"wonderful," "strange," "extraordinary." He admitted that he had been a fin spine booster, but in light of the new specimens, "this fine theory vanishes."

The bilateral symmetry of *Edestus* and *Helicoprion* structures had been widely accepted from the beginning, and this characteristic was also clear in *mirus*. Split longitudinally down the middle, the halves would be identical. Now, with *E. mirus* almost certainly proving that the structures belonged in the mouth, their bilateral symmetry likewise appeared to prove they were symphyseal, centered, structures. The *mirus* fossil, Hay said, showed that there was only one set of blades per animal.

"It is pleasant to credit Dr. C. R. Eastman with having in various papers advocated the idea that the tooth shafts of *Edestus* and related genera belong in the mouth," Hay wrote. "He has been disposed, however, to believe that there was a pair of them in one jaw or the other, probably the upper." Another fine theory out the window, at least in Hay's book.

Along with showing that the blades were in the mouth, the bit of preserved cranium also helped paleontologists determine the fossil's front from "hinder" end. "Here the opinion held by most writers is reversed," Hay wrote, including himself. That clue allowed them to determine that the tooth bases in *Edestus* pointed backward, toward the throat. Since they already knew *Helicoprion*'s tooth bases pointed in the opposite direction from *Edestus*'s, this meant that the tooth bases in the *Helicoprion* group must point forward, toward the front of the mouth. This scrap of intelligence was significant for the clues it gave as to which of the teeth were the oldest and which were newest, a detail that could help them begin to understand how the tooth structures grew. It also let them confidently call heads or tails on the fossils, instilling a new sense of motion and direction, should they want to picture those extinct creatures swimming in those mysterious ancient waters.

Hay leaped to the assumption that since *Edestus mirus* had an upper and a lower blade, *Helicoprion,* too, must have had whorls in both the upper and lower jaws—"like a pair of weapons resembling circular saws." Put that on a cocoa card! "Karpinsky's figure has seemed grotesque enough," Hay declared, "but it probably only tells half the story."

Karpinsky didn't venture any further *Helicoprion* reconstructions, although he did publish a paper in 1916 describing a new *Helicoprion* species, *H. clerci,* based on five additional fossil fragments from Krasnoufimsk. The world of paleo sharks was growing, but that wasn't the only world on Karpinsky's mind in those days. In March 1917, unrest over Russia's disastrous involvement in World War I and anger over chronic food shortages spurred the overthrow of Tsar Nicholas II. Less than a year later, the provisional government was upended by the Bolsheviks, led by Vladimir Lenin. Lenin agreed with his new minister of education that the Russian Academy of Sciences could help rebuild the country, and so should be politically and financially supported. Karpinsky was elected as the academy's first president, a post he held until his death, in 1936. History credits Karpinsky not only with establishing a solid role for the academy under the new regime, but also of saving a great deal of scientific equipment and many invaluable records during the months of turmoil and looting.

World War I also affected American members of the *Helicoprion* forum. Bashford Dean was commissioned as a major of ordnance and made the chairman of the Committee of Helmets and Body Armor of the National Research Council, helping produce a series of prototype helmets. C. R. Eastman was appointed to the War Trade Board, which oversaw imports and exports, rationed supplies, and kept strategic goods from enemy hands. The appointment might have been his final undoing, and the prolific paleoicthyologist died in 1918 at the age of fifty. An item in the *New York Times* read: "The body of a well-dressed man of middle age, who was found drowned at the end of the seawalk at Long Beach early Saturday morning, was identified yesterday as that of Dr. Charles Rochester Eastman, well known among American scientists. According to friends, he had been suffering from overwork while a member of the War Trade Board in Washington." Other obituaries attributed Eastman's fall from the boardwalk as an accident due to an attack of dizziness after suffering from influenza. It might have been true. He might really have tumbled into the water by accident. The flu pandemic of 1918–1919 killed more people than the war did, and the virus was said to strike people dead within hours of falling ill.

With Eastman's death, the forum had lost one of its most active voices and there was a steep drop-off in the conversation about curve-toothed paleo sharks. *Edestus mirus* had ostensibly settled a

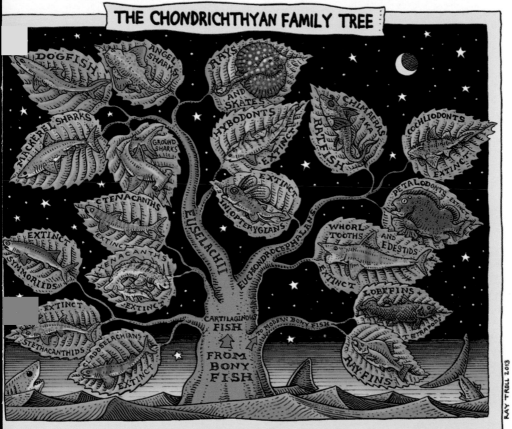

Chondrichthyans are a class of fish with skeletons of cartilage, rather than bone. Evolution split the class into two main subclasses: the Elasmobranchii and the Holocephali. The elasmobranch group contains the Euselachii, today's "true sharks" and their extinct relatives, while the Euchondrocephalan branch holds a diverse assemblage of species with holocephalan lineages, including *Helicoprion*.

Members of the *Helicoprion* research team gathered at Idaho State University in 2012 for a "Shark Summit." Left to right: Jason Ramsay, Cheryl Wilga, Alan Pradel, Robert Schlader, Jesse Pruitt, Leif Tapanila. *Photo by Ray Troll.*

Renowned chimera expert Dominique A. Didier affectionately known as the Ratfish Queen, with Troll ratfish art. *Photo by Ray Troll*

Dominique Didier and Ray Troll in 2011, with the ratfish Didier named for Troll *Hydrolagus trolli* ("Troll's chimera"), commonly known the pointy-nosed blue chime *Photo courtesy Ray Troll.*

...fe-size bust of *Helicoprion* by paleo-sculptor Gary Staab appears to burst through the wall in the 2013 ...Whorl Tooth Sharks of Idaho" exhibit at the Idaho Museum of Natural History. *Photo by Ray Troll.*

...e Pruitt fabricated an operating model of *Helicoprion*'s jaw for the museum exhibit. *Photo by Ray Troll.*

Leif Tapanila, Cheryl Wilga, Jason Ramsay, and Jesse Pruitt (left to right) scrutinize *Helicoprion* fossils for clues during the "Shark Summit" in Pocatello.

The process of mining phosphate generates rocky "overburden," like the rubble pile at this Idaho mine. Smooth, roundish concretions—some of which contain *Helicoprion* and other fossils—are sometimes salted among the rougher boulders.

Eminent American naturalist Joseph Leidy identified and named the *Edestus* genus. *Photo ca. 1853, used by permission of the Academy of Natural Sciences in Philadelphia: ANSP Archives Collection 9.*

American scientist Fanny Rysam Mulford Hitchcock was the first person to figure out that *Edestus* fossils were midline tooth structures. *Photo used by permission: University Archives and Records Center, University of Pennsylvania.*

Russian paleontologist Alexander Karpinsky identified and named the *Helicoprion* genus in an 1899 monograph that provoked animated international response. *Photo ca. 1897, from* Popular Science Monthly, *Volume 51.*

Edestus Davisii, H.Woodw.
Carboniferous: *Western Australia*.

(top left) Alexander Karpinsky's *Helicoprion bessonowi* type specimen. *Photo from Wikimedia Commons, by* Citron/ CC-BY-SA-3.0. *(top right)* Found in Western Australia, the world's first *Helicoprion* fossil was initially misidentified as *Edestus davisii*. Reproduction from *The Geological Magazine*, January 1886.

Canadian geologist W. F. Ferrier found the first American *Helicoprion* in Bear Lake County, Idaho around 1907. The fossil was initially misidentified as *Lissoprion*. *Photo courtesy Mineralogical Record Biographical Archive.*

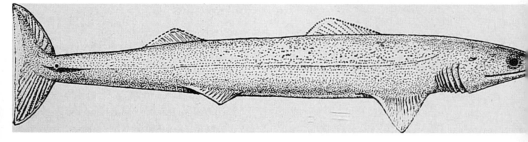

Cladoselache is among the best-known early sharks, thanks to well-preserved fossils from the Cleveland Shale. First appearing about 360-million-years ago, the shark grew up to six feet long. *Reproduction from Devonian Fishes of Iowa, by C. R. Eastman, 1908.*

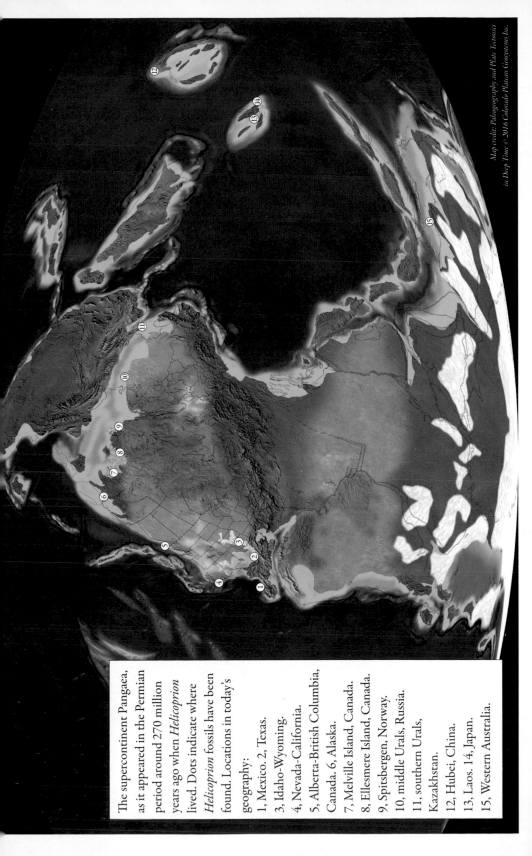

The supercontinent Pangaea, as it appeared in the Permian period around 270 million years ago when *Helicoprion* lived. Dots indicate where *Helicoprion* fossils have been found. Locations in today's geography:

1, Mexico. 2, Texas.
3, Idaho-Wyoming.
4, Nevada-California.
5, Alberta-British Columbia, Canada. 6, Alaska.
7, Melville Island, Canada.
8, Ellesmere Island, Canada.
9, Spitsbergen, Norway.
10, middle Urals, Russia.
11, southern Urals, Kazakhstan.
12, Hubei, China.
13, Laos. 14, Japan.
15, Western Australia.

Helicoprion may have used its tooth whorl as a sort of snail fork, to pull soft-bodied cephalopods from their shells. *Illustration by Jason Ramsay.*

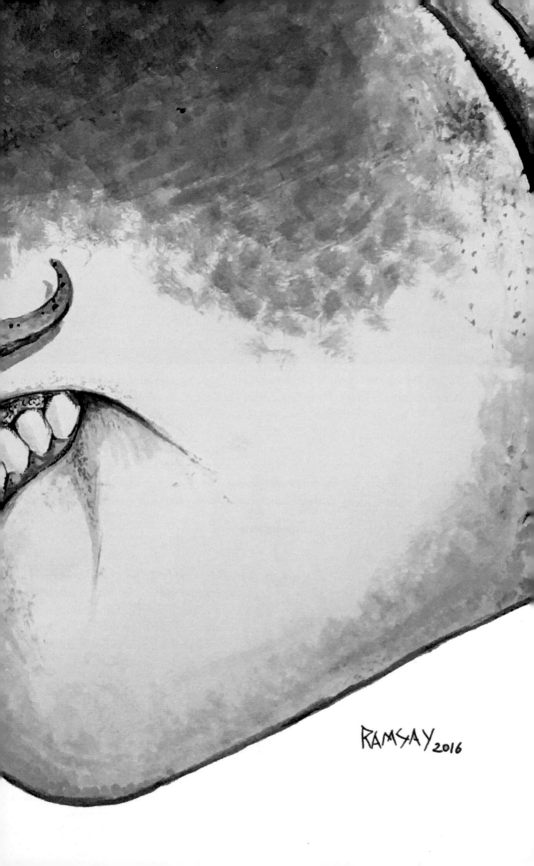

In 2009, Oleg Lebedev reconstructed *Helicoprion* with the tooth whorl at the front of an elongated lower jaw. *Illustration by Oleg Lebedev, used courtesy of the A.A. Borissiak Paleontological Institute of the Russian Academy of Sciences, Moscow, Russia.*

Evolution has given chondrichthyans many designs for attaching jaws to cranium, known as "jaw suspension." Different chondrichthyan subgroups—like sharks, rays, and ratfish—have different types of jaw suspension. Paleontologists use the feature to trace evolutionary lineage.

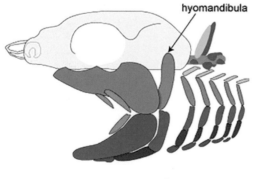

Most modern sharks have "hyostylic" jaw suspension, with the key attachment point coming from a "hyomandibular arch" positioned behind the jaw. This allows a shark to protrude its jaw forward when attacking prey.

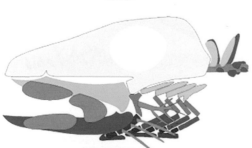

Many early chondrichthyans, including *Helicoprion*, did not have a hyomandibula. Instead, their upper jaw was suspended from the cranium by two muscle attachment points, an arrangement called "autodiastylic" jaw suspension.

Diagrams courtesy of Richard Lund and Eileen D. Gro

THE EUGENEODONTID SHARKS

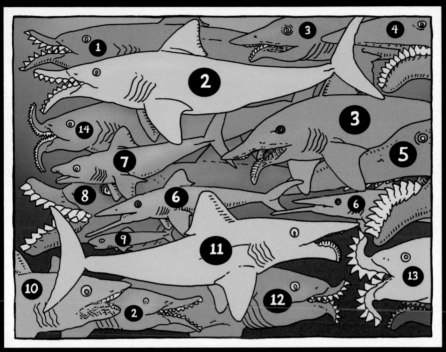

1. Edestus minor
2. Edestus heinrichi
3. Fadenia crenulata
4. Campyloprion ivanovi
5. Parahelicoprion clerci
6. Ornithoprion hertwigi
7. Caseodus eatoni
8. Edestus giganteus
9. Romerodus orodontus
10. Bobbodus schaefferi
11. Sarcoprion edax
12. Edestus mirus
13. Edestus newtoni
14. Toxoprion lecontei

Key to the eugeneodontid sharks, pages 12-13.

The Eugeneodontida were an order of Paleozoic sharks with curved or spiral tooth bases, including *Edestus*, *Helicoprion*, and a number of others.

The Eugeneodontid sharks, minus *Helicoprion*.
See page 11 for key. *Ray Troll © 2012.*

1. *Bandringa* Pennsylvanian, 2. *Heteropetalus* Mississippian, 3. *Edestus giganteus* Pennsylvanian,
4. *Iniopteryx rushlaui* Pennsylvanian, 5. *Promexyele peyeri* Pennsylvanian,
6. *Romerodus* Pennsylvanian 7. *Ornithoprion* Pennsylvanian, 8. *Helicoprion* Permian,
9. *Polysentor gorbairdi* Pennsylvanian 10. *Stethacanthus* Devonian-Pennsylvanian,
11. *Iniopera richardsoni* Pennsylvanian, 12. *Cobelodus aculeatus* Pennsylvanian,
13. *Orodus* Pennsylvanian, 14. *Sarcoprion* Permian, 15. *Harpagofututor volsellorhinus* Mississippian

Chondrichthyans arose and achieved impressive diversity during
the Paleozoic era (542-251 million years ago). *Ray Troll ©1994*

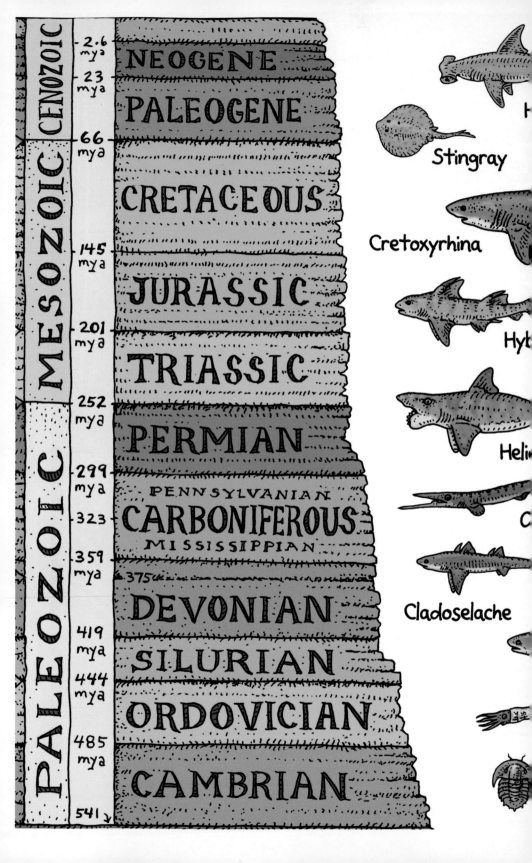

CENOZOIC		NEOGENE	2.6 mya
			23 mya
		PALEOGENE	
MESOZOIC			66 mya
		CRETACEOUS	
			145 mya
		JURASSIC	
			201 mya
		TRIASSIC	
PALEOZOIC		PERMIAN	252 mya
			299 mya
	PENNSYLVANIAN	CARBONIFEROUS	323
	MISSISSIPPIAN		359 mya
		DEVONIAN	375
			419 mya
		SILURIAN	
			444 mya
		ORDOVICIAN	485 mya
		CAMBRIAN	
			541 y

Stingray

Cretoxyrhina

Hyb

Heli

Cladoselache

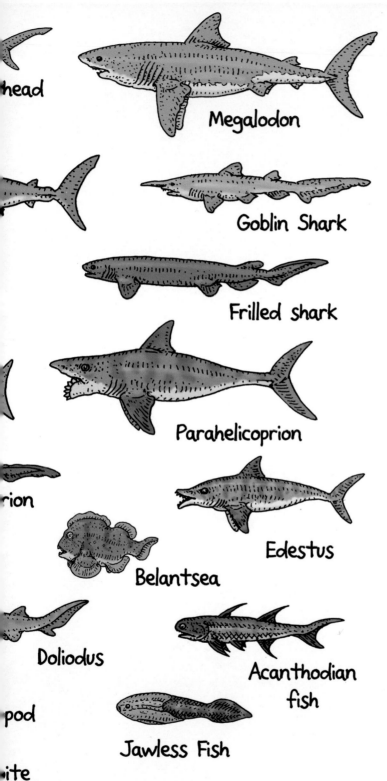

head

Megalodon

Goblin Shark

Frilled shark

Parahelicoprion

rion

Edestus

Belantsea

Doliodus

Acanthodian
fish

pod

Jawless Fish

ite

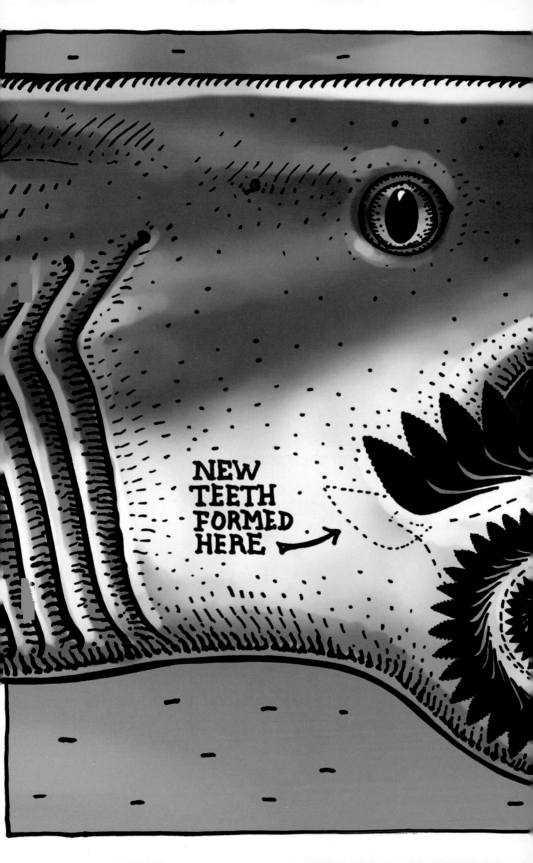

New tooth crowns erupted from a "tooth pit" near the back of the mouth, slowly pushing older crowns forward. *Ray Troll © 2012*

HELICOPRION
hēel-ih-cōe-pry´on or
hĕll-ih-cōe-pry´on
~ means ~
"spiral saw"

In the century between Karpinsky's first snout-whorl sketch *(top left, page 20)* and Ray Troll's best-current-research illustration *(lower right, page 21)*, scientists from around the world imagined where *Helicoprion*'s whorl might have fit on the animal.

Karpinsky 1899

Karpinsky 1902

Karpinsky 1911

Van Den Berg 1953

Eaton 1962

John Long 1995 ↗

Lebedev 2009 ↗

Top left sketch by Karpinsky, all other art by Ray Troll, from scientists' original drawings.

The images on pages 22-24 are target art for augmented reality models developed by Jesse Pruitt and the Idaho Virtualization Laboratory. See page iv for app information."

Idaho number

Idaho no. 4, the fossil that helped crack the *Helicoprion* code. Stereo stacked image by Nick Holme
(The augmented reality model of Idaho no. 4 lets you zoom in on the fossil.)

(opposite) CT scans from Idaho
4 allowed the team to reconst
Helicoprion's upper and lower ja
Montage illustration by Jesse Pruitt. (
augmented reality model shows the j
opening and closing over the whe

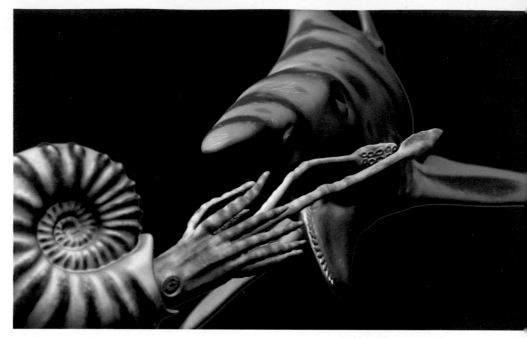

Helicoprion chasing an ammonite, its likely preferred prey. Illustration by Jesse Pruitt. (The augmented reality model is a full-body model of the *Helicoprion* and ammonite.)

Rainerichthys zanger

Rainericthys zangerli belonged to an order of chimera-like chondrichthyans, the Iniopterygiformes (or "iniops"), that lived around 345-280 million years ago. The illustration and augmented reality model are based on the head fossil (upper left) and skull reconstruction (upper right). Fossil and skull diagram courtesy of Richard Lund, illustration by Jesse Pruitt. (The augmented reality model is a full-body model of *Rainericthys*.)

lot of questions anyway, and helmets and body armor needed to be built for the boys in the trenches. After the early-twentieth-century feeding frenzy, the sharks would lie still until the dawning of the next World War, when a freshet of new fossils in America and Australia emerged to push *Helicoprion* into the scientific spotlight once again.

8

SIGNS OF LIFE

A ten-pound sledge is a very handy thing.
—Walter Youngquist, retired paleontologist and teacher,
reminiscing in 2014 at age ninety-three

AFTER W. F. FERRIER SENT HIS CONSEQUENTIAL CRATES OF PHOSPHORIA FOSSILS TO Oliver Perry Hay, twenty-five years passed before any more *helicoprions* were found in the United States. The dry spell was broken in 1929 when Elbert A. Stewart discovered a whorl in a lens of volcanic tuff six thousand feet up in the Humboldt Range of northwestern Nevada. The specimen was a mold, an indentation preserving external details, almost five inches in diameter, the size of a big grapefruit. It had two and a half volutions, and held impressions of ninety-five finely serrated teeth over an inch tall. Not a single fossil had previously been found in that whole "dominantly eruptive" formation, so might as well make it a good one. Unfortunately Stewart couldn't toast his surprising find, at least legally, because America was still in its own dry spell of Prohibition.

Another *Helicoprion* whorl was collected in 1931 by J. H. Menke, from a glacial moraine in California's northern Sierra Nevada. Like the Nevada fossil and Mr. Davis's Australian fragment, the California specimen was also a mold, but in this case portions of the teeth and root-base were still remarkably adhered to the rock. The spiral was over six and a half inches in diameter, the size of a large cantaloupe, with three and a quarter volutions and evidence of 109 teeth up to an inch and a half tall. Menke couldn't toast his incredible find either.

Harry E. Wheeler, geology professor at the University of Nevada's Mackay School of Mines, described the finds in 1939, so he *could* drink to the swell new American sharks. He called Stewart's fossil *H. nevadensis,* and gave Menke's the name *H. sierrensis.* In his paper on the whorls, Wheeler wrote, "These discoveries are significant not only because *Helicoprion* proves important as a stratigraphic index, but also because they throw new light on the paleogeography and sequence of events in western North America." Western stratigraphy was still being sorted out, and geologists had considered rocks where the whorls were found to be of Mesozoic age, the time of dinosaurs and pterosaurs. The *helicoprions* provided evidence that the strata were from the much older Permian. These new American fossils, as well as other *helicoprions* cropping up around the world, were consistently being found in the same relatively small horizon of Permian strata encompassing a span of roughly ten million years, from about 270 million to about 260 million years ago. Consequently, Wheeler wrote, *Helicoprion* should be regarded as a "zone fossil," or indicator species, for the time just before the middle of the Permian. Wheeler noted the big blank line, however, next to the Australian fossil that started it all. Mr. Davis's Australian

Helicoprion had not had its time card officially punched because it wasn't found by geologists in situ, in place. Wheeler wasn't the only one thinking about that missing piece of information. Australian geologists especially were stressing the importance of pinning down the fossil's accurate age. In 1931, Aussie scientists T. W. E. David and C. A. Sussmilch wrote, "This occurrence of *Helicoprion* (?) is obviously of vital importance for the whole paleontology and stratigraphy of the 'Permo-Carboniferous' rocks of Australia, and every effort should be made to discover the exact horizon of its occurrence and to ascertain whether it had a spiral."

Every effort is good, but in field paleontology, sometimes what you really need is luck.

Shirley Gooch could have been watching a bright flock of wild budgerigars flying over the arid Australian scrubland and stepped right over it. Or she could have put the interesting fossil in a box and forgotten about it. But she didn't. In 1937, Gooch picked up a fossil of five teeth with sharply bent bases in the bed of the Minilya River near Wandagee Homestead. This was about 120 miles north of where Mr. Davis found the world's first *Helicoprion* some fifty years earlier. We don't know how old Shirley Gooch was, or what she was doing that day, but it's reasonable to guess she was the granddaughter of G. J. Gooch, who established Wandagee Station in 1880. Sheep were Western Australia's new gold, and the Gooches were known for their high-quality merinos. Shirley's fossil somehow made its way to Perth, where it came to the attention of paleontologist Curt Teichert, then at the University of Western Australia. In 1939, Teichert led a small party to the same area, where they found a fully spiraled *Helicoprion* a mile north of Wandagee Hill. Finally, hallelujah, Australia had its scientifically meaningful in situ specimen. Actually it was double hallelujah: "Mr. H. Coley picked up an external mould of a *Helicoprion*," Teichert wrote, "and a few yards from this place Mr. E. P. Utting found part of the mould of the opposite side of the same specimen."

The fossil was as impressive as the find itself, with nearly three full volutions. It must have been an exciting moment for Teichert, the only professional paleontologist working in Western Australia at the time, and one of just six on the whole continent. Teichert had only recently come to Australia. Born and largely educated in Königsberg, Prussia, then part of the German Empire, Teichert received his PhD in 1928. That same year he happily married Gertrud "Trude" Kaufmann, daughter of a physics professor. A couple of years later, Teichert received a Rockefeller Fellowship to study early cephalopods at the Smithsonian's National Museum of Natural History, an expertise that led to a fifteen-month Danish expedition to Greenland. By the time Teichert returned to Germany in 1933, the Nazis were in control. When university authorities told him to divorce Trude because of her "non-Aryan" ancestry, the couple hastily left for Copenhagen. They remained deeply devoted to each other throughout a thriving, productive, and fruitful sixty-six-year marriage—although they were lucky to eke out a living during those first few years in exile. They sheltered in Copenhagen until 1937, when another Rockefeller Grant, this one for resettling displaced German scientists, enabled Teichert to take a faculty position in the isolated city of Perth, at the small University of Western Australia, in the tiny department of geology. He had little money for research but took a big step in overcoming that hurdle when he discovered beer was a valid currency for access, boat transportation, and fraternity with the locals.

The Teicherts escaped the Nazis but not the hysteria of the times, and when World War II broke out in 1939, the couple was briefly interned by Australian authorities. It's not clear if this was before or after the May field trip to Wandagee Hill. The Teicherts were "offered the opportunity" to return to Germany in return for Australian prisoners, and the government seemed surprised when the offer was declined. They were still German nationals, so upon their release they were subject to certain travel restrictions, but Teichert was able to continue his teaching, research, and publication.

Teichert published his paper on the *Helicoprion* finds in 1940. With the new Wandagee whorl—the full spiral that David and Sussmilch had been hoping for—he was able to confirm Karpinsky's 1899 proposition that Mr. Davis's fossil was truly a *Helicoprion*. Not *Edestus* as Henry Woodward had originally said, not *Campyloprion* as Eastman asserted, and not *Toxoprion* as Hay proposed. This confirmation was an important pushpin on the map of *Helicoprion*'s global distribution, and the in situ find advanced the knowledge of Western Australia's stratigraphy by virtually putting a "you are here in time" sticker on the Wandagee strata.

In his paper, Teichert paid tribute to Karpinsky's elaborate description of *Helicoprion*, which he called "still unsurpassed." He noted that Karpinsky had correctly understood the whorl as a solid structure—all those "teeth" were really one tooth with multiple crowns sprouting from a single continuous root. Given this unique feature, Teichert agreed with Karpinsky's 1911 decision to erect a separate family for the shark, Helicoprionidae, in contrast to other paleontologists who had continued to include *Helicoprion* in the Edestidae family. (Names ending in *ae* signify a family.)

Although no vocal advocates of the fin spine theory were left standing, Teichert acknowledged that the persistent, perplexing, and confounding question of where the tooth whorl fit on the animal remained a big unknown. Karpinsky had died a few years earlier (with his body interred in the Kremlin Wall Necropolis), still of the opinion that the whorl was a spiral of teeth existing for the most part outside the mouth. The eminent Russian had remained open, however, on the question of whether the spiral had been present in both jaws or just one. And if just one, which one?

Teichert, content for the moment with all the other morsels of information contained in the new Australian fossils, let the question of whorl placement ride. "Even today, no answer to these questions can be given, and no better statement of the case can be given than those presented by Karpinsky." The ultimate answer to this most persistent, long-standing question was more than seventy years in the future—but the extraordinary fossil that would provide the key was about to see the light of day back in Idaho.

It was hot in the sagebrush hills of southeastern Idaho, especially hiking around with a ten-pound sledgehammer. The tool might have seemed like overkill, considering the young paleontologist hefting it, Walter Youngquist, was looking for microfossils. It was July 1949, and Youngquist was on the hunt for toothlike "elements" left behind by conodonts, poorly understood soft-bodied creatures dating back to the Cambrian. Conodont elements are tiny—several can fit on the head of a dressmaker's pin—and take an array of spiky, knobby, comblike shapes that come into beautiful focus under a microscope. In the way *Helicoprion* raised attention, controversy, and consternation at the beginning of the twentieth century, the enigmatic little conodont was doing the same mid-century, ever since someone noticed it was associated with oil-bearing rocks. Scientists were arguing, sometimes heatedly, about what the organisms were (fish ancestors? marine worms? plants?), and whether the spiked elements were evolution's earliest teeth or something else. A conodont rush was on, to find them, then figure them out well enough to predict where to find more.

After serving on a navy submarine chaser during World War II, Youngquist had completed graduate school at the University of Iowa and joined the ranks of newly minted scientists. He found his first teaching job at the University of Idaho in Moscow. That summer he had a thousand-dollar grant to conduct the fieldwork of his choice, and he chose to look for conodonts. As far as he knew they'd never been found in the Phosphoria, but maybe because no one had thought to look for them there. One of his stops was the Waterloo Mine in Bear Lake County—Idaho's first phosphate mine, opened in 1907 almost certainly with the engineering help of W. F. Ferrier. That put Youngquist poking around in the same county, maybe near the very spot, that Ferrier found the first American *Helicoprion*. The significance was lost to Youngquist, who wouldn't have recognized a *Helicoprion* if one bit him on the ankle.

Since Ferrier's day, the Waterloo mine had been worked hard as an underground operation until the 1920s, then closed from 1929 until 1945, when it was reopened and worked intermittently as a surface mine. The mine was shut down when Youngquist and his field assistant Jerry Haegele showed up looking for their microfossils. Those were the days of casual access, and Youngquist and Haegele helped themselves to the site. Youngquist was wandering around, scrutinizing slabs of shale through his hand lens, when he noticed a pile of concretions at the base of the pit. Concretions are what they sound like, concrete-hard formations of varying sizes, usually round or oval, that develop within strata and are harder than the surrounding rock. They form through a chemical process, often around a nucleus of organic material like a leaf or shell. Concretions are a pain in the neck to miners, who toss them aside into waste piles, but paleontologists love them for the Cracker Jack surprises that may, or more likely may not, be in them. You can break open a lot of concretions and come up empty. But sometimes, there is a prize. So Youngquist put his hand lens in his pocket, located the bedding plane on one of the concretions in the pile, swung his sledge and whacked. The concretion popped cleanly open to reveal a *Helicoprion* whorl neatly nested inside. Youngquist had no idea what it was. He eyed another concretion, swung the sledge, whacked and scored again. By the time he put down the sledge he had seven *Helicoprion* fossils. It's hard to say for sure, but until that moment, the world's scientific inventory of *Helicoprion* fossils was likely around fifteen to twenty specimens. Youngquist still holds the record for a single discovery.

When they were ready to go, Youngquist and Haegele loaded the *Helicoprion* fossils and other finds in their car. In the bonanza was one that didn't look much different from the others there at the bottom of the pit, but which years later would play a pivotal role in unlocking the mystery of the whorl. As the heavily loaded car pulled out onto the empty Idaho road, a litter of fossil counterparts receded in the rearview mirror. In looking for the smallest, possibly earliest known precursor of vertebrate teeth, Youngquist had stumbled on one of evolution's oddest, most elaborate tooth experiments, writ large.

Back in Moscow, Youngquist dissolved his shale samples and discovered they did hold conodont elements, the first from the Phosphoria. He also identified his unparalleled bounty of tooth whorls to be *Helicoprion* fossils. Now the whorls had real value to him, for what they meant in relation to his conodont investigations. As recognized zone fossils, the *helicoprions* confirmed that Youngquist's Phosphoria conodonts dated to the Permian—the youngest yet to be systematically described. In a 1951 paper on the findings, Youngquist cited the "incidental discovery" of the *helicoprions* as one means of securing the age of his conodonts, since *Helicoprion* was indicative of the mid Permian period. The next summer Youngquist would find conodont elements in Idaho's Triassic rocks, a bombshell of a discovery upending long-held assumptions about how long conodonts had existed on the planet. Conodonts were initially thought to have died out at the close of the Paleozoic era, which ended around 250 million years ago with the Permian mass extinction. But Youngquist's findings showed that they lived on for perhaps nearly another fifty million years, into the late Triassic.

Despite his wave-making work, Youngquist wasn't happy in Idaho. By 1953, four out of five geology graduates were making good livings as petroleum geologists, and after three years of teaching in Idaho, Youngquist had only managed to save eighty dollars. So he chucked the conodonts and went to Peru in the employ of Jersey Standard. The University of Idaho didn't have a museum, so before he left for South America, Youngquist sent the *Helicoprion* fossils to his Iowa alma mater. William Furnish, head of Iowa's paleontology department, took possession of the whorls.

––––––––––

Around the time Youngquist was heading for Peru, Danish pale-ontologist Eigil Nielsen described an odd new prion that had been found on a mountain slope in Greenland. *Sarcoprion* (*sarco*, meaning "flesh,"*prion*, "saw"—flesh saw) had a swordfish-like ros-trum and a relatively small half-whorl at the tip of the lower jaw. The first *Sarcoprion* fossil was discovered in 1932 in a large, broken concretion that was missing some of its parts. Incredibly, Nielsen found two important fragments from the concretion on a return trip in 1937. In all, he made three lengthy expeditions to Greenland in the 1930s, gaining a reputation as an outstanding mountaineer, fossil hunter, and dog musher. Greenland had been under Danish rule since 1397, with colonies established in 1721, and systematic scientific exploration beginning in 1875. In the 1920s and 1930s, Norway challenged Denmark's sovereignty over the polar island, spurring even more activity by the Danes, including paleontolog-ical expeditions. It appears Nielsen would have been happy to stay in Greenland his entire career (his superiors called him back against his wishes in 1939), but eventually he settled in as head of the Department of Vertebrate Paleontology at the Mineralogical and Geological Institute of the University of Copenhagen. He finally found time to describe the *Sarcoprion* in 1952.

Nielsen had a student named Svend Erik Bendix-Almgreen, his eventual successor, who shared his intrigue with whorl-toothed sharks and Greenland. Bendix-Almgreen had himself led two small Greenland expeditions in 1958 and 1959, and afterward wrote a paper in Danish on "The Anatomy of Permian Edestids from East Greenland." Probably while he was working on that paper, maybe at

Nielsen's suggestion, Bendix-Almgreen decided to look closely at Idaho's Phosphoria *helicoprions*. It's unclear how he knew about Youngquist's fossils, or how he tracked them to Iowa. But in the fall of 1961, William Furnish handed over Youngquist's Idaho *helicoprions* along with a few others, ten specimens in all, to the twenty-nine-year-old Dane. All but one of the fossils had been collected at the Waterloo Mine, with the outlier found at the Gay Mine not too far away. (Although they had been housed in Iowa all that time, the fossils were noted as property of the University of Idaho and Idaho State College.) In addition to the fossils, Bendix-Almgreen received documents accounting for seventeen other previously undescribed Idaho *helicoprions*. Ten were stored in collections at the US Geological Society and Smithsonian National Museum of Natural History in Washington, with the others in private possession.

Bendix-Almgreen studied the material for about four years, and published his research in 1966. "New Investigations on *Helicoprion* from the Phosphoria Formation of South-East Idaho, U.S.A." was the most thorough and illuminating paper on *Helicoprion* since Karpinsky's original 1899 treatise, logging in at fifty-two pages with abundant figures and plates.

A decade earlier, Russian paleoicthyologist Dmitry Vladimirovich Obruchev, son of famed paleontologist and science fiction writer Vladimir Afanasyevich Obruchev, had written an overview of Karpinsky's *Helicoprion* work with illustrations summarizing the historical ideas about whorl placement, including the original snout whorl, the tail whorl, the double dorsal whorls, and the Belgian whorls. For his own contribution, Obruchev improved on Karpinsky's snout whorl by illustrating it with all but the perimeter of teeth covered with cartilage to make it look solid, presumably to give it more punch as the weapon they thought it was. Like so many before him, Obruchev was on the right track (with the encasing cartilage), but still on the wrong train, locating the whorl on the snout. "One may hope," Obruchev wrote, "that new finds will show how near such a reconstruction is to the truth."

How near, or how far. Bendix-Almgreen's research was a quantum leap forward in the understanding of *Helicoprion*. One fossil in particular—one of Youngquist's incidental whorls—held new, vital information. There was much to look at on "Idaho no. 4," as Bendix-Almgreen labeled it, with its three and three-quarter volutions and 118 teeth, some of which retained traces of enamel. The tooth impressions were incredible and the spiral was entrancing, with a little fishhook curve clearly outlined at the center, but Bendix-Almgreen's attention was arrested by a far more subtle feature. To an untrained eye, it would have simply looked like patches of differently textured rock scattered around the face of the fossil, something a casual observer would look right past, like lichen on a headstone. But Bendix-Almgreen realized he was seeing preserved cartilage. The patches were two different textures, or "complexities," as he called them, of "prismatic calcifications." In other words, Idaho no. 4 contained fossilized tessellated cartilage—a phenomenally rare circumstance and thus far unprecedented in *Helicoprion* fossils.

Tessellated cartilage is one of the core identifying characteristics of sharks and their kin—and a very slick evolutionary maneuver by class Chondrichthyes. Chondrichthyans, by definition, have skeletons made of cartilage rather than bone. Cartilage is about half as dense as bone, so having a cartilaginous skeleton significantly reduces a shark's body mass, enabling even the largest individuals to move economically and efficiently through the water. But while pure cartilage works beautifully for functions that benefit from lightness and flexibility, like swimming, having a whole skeleton of the rubbery tissue wouldn't support a large-bodied, aggressive predator that sits high on the food chain. It would be hard to take down a sea lion or sea turtle with a jaw made from basic cartilage. So sharks figured out a way to have their cake, or seal, and eat it too, by evolving a special type of cartilage in which tiny mineralized plates called "tesserae" form over parts of the cartilage in a mosaic held together by a mesh of connective tissue. (The word is taken from the name for the small tiles that ancient Romans used to create their mosaics.) Tessellation significantly strengthens cartilage while letting it remain flexible and relatively light. Most of a shark's skeleton is tessellated to some degree, with extra layering in places like the jaw and cranium. A shark continually adds new tesserae as it ages and gets larger, which effectively means its entire life, since sharks are thought to grow until they die. Interestingly, scientists have long thought cartilaginous skeletons were the "primitive" condition in sharks, but recent discoveries and research by Australian paleontologist John Long revealed that the earliest fish identified as being sharklike may have had much more bone in their skeletons than today's sharks. Which means, Long says, the evolution of modern sharks appears to have been driven by a *loss* of bone. With that reveal, we don't have to imagine sharks as the evolutionarily inferiors of the bony "higher" vertebrates (like the ones sharks eat). The chondrichthyans saw a fine future for themselves down a more flexible path.

According to Bendix-Almgreen, the tessellated cartilage visible on the face of Idaho no. 4 represented bits of jaw and cranium preserved "to some degree in their original positions." Bendix-Almgreen also detected thin lines on the broken edge of the rock that he interpreted as cross sections of the upper and lower jaws. With that, the whorl clicked into place and Bendix-Almgreen had his theory on where it fit.

"In the past, some confusion has prevailed over the question of the original location of the tooth spiral," Bendix-Almgreen wrote, in a generous reflection on the preceding six decades. "The most adequate previous explanation was given by Karpinsky, although it was viewed with great skepticism by many authors." But now at last, Bendix-Almgreen was ready to give the blind men their elephant. It was a dramatic moment, undramatically expressed. "Our present extended knowledge enables us to state with certainty," he wrote, "that the tooth-spiral is, in fact, situated in the symphysis of the lower jaw and that no similar organ was developed in the upper jaw."

Using techniques he learned from Nielsen, Bendix-Almgreen set about triangulating the scraps of evidence into a diagram of the whorl placement as best he could, without being able to peer inside the rock to see cartilage that might be buried there. He drew the tooth spiral half-imbedded at the front of the lower jaw, and noted that the two halves of the jaw must have formed a cavity to accommodate the inner, older, parts of the whorl.

Confirming the previous judgments of Karpinsky, Eastman, and Teichert, Bendix-Almgreen described the whorl as having a solid, coiled compound root-base, developed through a complete fusion of each new tooth—or rather each new crown. To avoid confusion and support the whorl as one unified structure (one tooth), Bendix-Almgreen replaced the terms "teeth" and "shaft," with "tooth crowns" and "compound root."

To decipher the whorl's development and growth, Bendix-Almgreen introduced two more key components that he sleuthed from Idaho no. 4, in addition to the preserved cartilage: the tooth pit and the juvenile arch. The tooth pit, located at the rear of the whorl, toward the throat, was where new teeth formed and emerged. As we know, modern sharks produce an endless conveyor belt of new teeth that begin deep in the soft tissues of the mouth and are essentially built from the outside in. So in *Helicoprion*, we can visualize an enameloid crown being formed and filled by root material in the tooth pit, then being pushed forward and out into the mouth as the root—*Helicoprion*'s conveyor mechanism—continued to grow.

And what kept the tooth growing in a tight spiral, instead of shedding its crowns or snaking out of the mouth? That would have been the juvenile tooth arch, the little fishhook curve Bendix-Almgreen saw at the center of Idaho no. 4. The tooth arch is an evolutionary novelty at the literal heart of *Helicoprion*'s whorl, like the eye of a tornado, hub of an unfurling fern, or spiraled center of an ammonite shell. The juvenile tooth arch set the reeling action in motion and established the geometry of whorl growth. The tooth arch was an enamel-covered rod with a curved end that altogether formed about a fifth of the first volution. When root growth pushed it from behind, it began to curl in on itself instead of moving linearly. Other prions may or may not have had something like a juvenile tooth arch, but the particular curve that set up a fully spiraled geometry was apparently unique to *Helicoprion*. The first tooth crown appeared at the tail end of the juvenile tooth arch. The spiral's first volution held mostly "baby teeth" that all looked about the same. By the second volution, tooth crowns had matured into strong, pointed cutting blades with lightly serrated cutting edges. By the third volution, tooth crowns were getting progressively larger and larger, with clearly distinguishable cutting blades, midsection, and narrowed base.

In Bendix-Almgreen's reconstruction, the tip of the lower jaw had a short, prong-like strut supporting the leading edge of the whorl, with about half a dozen or so tooth crowns exposed at the front of the chin before they were absorbed into the cavity as the whorl notched forward in its growth.

"The reasons for the development of such an unusual organ as the *Helicoprion* tooth spiral are certainly obscure," wrote Bendix-Almgreen, "especially since in all the previously described species no observations of evidence of actual use have been reported." The lack of strong wear marks and the whorl's position in the mouth allowed him to rule out some existing speculations: the whorl was not a crushing organ, so *Helicoprion* wasn't rooting brachiopods out of the substrate to eat. And it wasn't a weapon, since it was inside the mouth. Bendix-Almgreen had detected faint wear marks on one of the Idaho specimens and some of the Russian fossils though, and had identified what he took to be small pavement-type teeth associated with the upper jaw in Idaho no. 4, so he suggested *Helicoprion* probably used the whorl to cut, and maybe tear, soft prey.

The well-preserved Idaho no. 4 also had a distinct rounded knob about the size of a quarter in front of the spiral. Bendix-Almgreen called it a "tuberosity" and thought it might be a point of attachment for lower jaw ligaments. The more modern term for the knob is "process," a projection on cartilage or bone that often anchors muscles and ligaments. He thought he could make out a "foramen" too, an opening in cartilage or bone that provides a passageway for veins and nerves.

Scrutinizing the cartilage visible on the fossil, Bendix-Almgreen surmised certain things about the way the upper jaw attached to the skull, what scientists call "jaw suspension." Feel free to picture suspenders. Suspenders sling around your shoulders to hold your pants up, and tendons and joints create a suspender system to hold a shark's jaw to its cranium. (Cranium, neurocranium, and brain case are somewhat interchangeable and refer to the part of the skull that houses the brain. The skull is the head's entire bony/cartilaginous structure, including jaws.) Depending on how you count, there are up to half a dozen different types of jaw suspension, and it's one of the most important factors in how chondrichthyans are grouped taxonomically, and how paleontologists trace the major lineages of chondrichthyan evolution. For now, just know that jaw suspension is among the primary features that set "true" sharks apart from holocephalans, which today we know as ratfish, and whose evolutionary lineage is long, diverse, and infinitely interesting. In holocephalans, the upper jaw (technically the palatoquadrate), is one continuous part of the cranium. It's fused, as scientists say. No suspenders required. This is at the very core of the name: *holos,* "whole," and *kephalos,* "head." Sharks, skates, and rays have several styles of jaw suspension between them, in which the upper and lower jaws are not rigidly fixed to the cranium, but are attached by various arrangements of elastic ligaments and other elements. Such a system allows some sharks, very notably goblin sharks and great whites, to project their jaws forward up to several inches when snatching prey.

Certain members of the de facto *Helicoprion* forum had suggested the creature might be in the holocephalan line of chondrichthyan evolution because symphyseal dentition was a feature of that side (mostly as fused, grinding plates), and there had been no previous clues to jaw suspension. Because of what Bendix-Almgreen deduced about the upper jaw, however, he thought *Helicoprion* was more closely aligned with the shark side. In explanation, he suggested that parallel evolution and specialization of the teeth in the symphyseal region of the lower jaw had taken place within different elasmobranch lines during Carboniferous and Permian times.

Whichever route it took up through time, *Helicoprion* was starting to show real signs of life—cartilage and processes and foramens, conjuring jaws and ligaments and nerves. The de facto forum had waited a long time for that. The original members were all dead by then of course—Karpinsky, Eastman, A. S. Woodward, Dean, Hay, and the rest—but with the apparent evidence in Idaho no. 4, maybe they could rest in greater peace.

Among the many illustrations in his paper, Bendix-Almgreen penned the iconic black-and-white tooth whorl graphic, based on Idaho no. 4. Stare at it long enough, and it starts to pulse on the page. When he was finished with them, Bendix-Almgreen sent the fossils back to Idaho, to what was then the Idaho State University Museum in Pocatello (not to be confused with the University of Idaho in Moscow). The museum's holdings were strewn around the campus until 1977, when the museum began a new focus on natural history and consolidated its collections in the old library building. Who knows how many times Idaho no. 4 was shuffled on a cart from one building to the next as it waited for the next generation de facto forum to feel its pulse.

9

THE ART OF OBSESSION

The artist and the scientist bring out of the dark void, like the mysterious universe itself, the unique, the strange, the unexpected.
—Loren Eiseley, *The Mind as Nature*, 1962

WHEN RAY TROLL LEFT THE NATURAL HISTORY MUSEUM OF LOS ANGELES COUNTY WITH his writer friend Brad Matsen that January day in 1993, the artist's head was spinning. The big tooth whorl J. D. Stewart showed them in the basement was mind-boggling. Troll had seen a parade of remarkable fossils during his and Matsen's research for their book-in-progress, *Planet Ocean*, but this one . . . This stony spiral was his stranger on the train, pushing out every other thought.

Troll didn't know a thing about Paleozoic sharks. He had never heard of *Helicoprion* or Alexander Karpinsky or symphyseal dentition. He didn't know that the fossil whorl, the museum's only one, came from Bear Lake County, Idaho. He only knew that he felt suddenly, thrillingly consumed. Holed up in a motel room that night with his sketchbook, Troll drew his first *Helicoprion* with nothing to go on but unbridled ardor and imagination. Troll sketched the whorl tucked up under *Helicoprion*'s chin and endowed it with the power to whip out like a New Year's party horn, indicated by energetic dashed lines. He gave his *Helicoprion* the tiger striping of a leopard shark and the long nose of a blue shark, and exuberantly labeled it the "Snap Towel, Party Favor, Table Saw, Whorl Shark."

Matsen and Troll headed up the coast the next day, capping off their drive with happy hour at the Elkhorn Yacht Club on Monterey Bay. Just south of San Francisco, Monterey Bay is one of the largest protected marine areas in the world, providing relatively safe harbor to all manner of sea life, including otters, dolphins, whales, and sixteen species of shark, from the small gray smoothhound to the enormous great white. Established in 1947, the neighborly club offered its own form of refuge to boaters and fishermen of all stripes. It was a friendly place that might smell like tacos on Tuesday night and clams on Friday. Matsen once told Troll if you can explain a concept to your bartender you're a good writer, so Troll set out to explain *Helicoprion* to the yacht club bartender and whoever else would listen. Quickly coming up against the fact that he didn't actually know anything about the beast, he settled for trying to generate curiosity and enthusiasm through brisk sketches. Running out of cocktail napkins, Troll commandeered the white board behind the bar, rubbing out the night's drink specials and illustrating a whorl-toothed shark lunging toward unseen prey.

"I was diving for abalone," came a German-accented voice a few stools down. On the California coast, sharks focus people's attention and open the tap on stories. The man at the bar had been scuba diving for abalone in Monterey Bay, using an underwater scooter. He was beginning to surface when he saw a dark form coming at him from the side. It veered past, he said, then a huge tail washed in front of his face and faded from sight. "I thought, 'Wow, that was a big shark! And I didn't have to pay any money for it!'" Shark tours were becoming popular, with people paying big bucks to be submerged in a protective cage for close shark encounters. The man was still about twenty feet from the surface, he said, when he looked down into a giant, open mouth lined with teeth, coming straight at him from the gloom. The shark hit him, bit him, and moved off. The man was bleeding profusely when he surfaced, but intact. The shark bit the scooter too, which might be why it backed off. It was almost certainly a great white. Putting down his drink and lifting up his shirt, the man revealed a swath of scars running down his side and disappearing below his belt line—almost to his knees, he said.

It's hard to top a story like that, especially when all you have to offer is a five-syllable scientific name and a cockamamie set of chop-saw teeth. Troll lost his audience as people drifted back into conversations about deck finish, upcoming races, and what they wanted to eat. Troll was left alone with his mythical beast, which followed him home to Ketchikan, sculling behind him up the steep steps to his hillside studio.

Troll's education had been in fine art and printmaking, with his whimsical, surrealist fish-themed prints on T-shirts and posters paying the bills. There was DIVE BAR, with fish bartenders serving fishy-looking patrons, TORTURED SOLE, and so on. The artwork Troll was doing for *Planet Ocean* was his first focused plunge into actual paleoart—the illustration of extinct creatures.

The field of paleoart came into its own in late-nineteenth-century New York City in a fractious but productive relationship between artist Charles Knight and Henry Fairfield Osborn, a paleontologist and the president of the American Museum of Natural History. Knight, who began his career as a commercial artist and freelance illustrator, was intensely interested in nature and animals. In 1894, his frequent sketching visits to the American Museum led a museum scientist to ask him to paint a warthog-like animal, *Entelodon*, whose fossil bones were on display. Knight used what he knew about porcine anatomy to stand the beast up, then applied his imagination to flesh out the rest. The museum and the museum-going public were enthralled. From that beginning, Knight broke new ground in the realistic, artistic illustration of fossil animals. He was the first, in 1897, to portray dinosaurs in vivid action with his "Leaping Laelaps," showing two tyrannosaurids in dynamic combat. Some curators complained that Knight's work was more artistic than scientific, and argued that his lack of a formal science education disqualified him from creating such fully imagined illustrations. Knight agreed that his murals were often works of art, but he also insisted he knew as much about paleontology as the curators did.

In the venerable Knight tradition, Troll was illustrating the prehistoric creatures in *Planet Ocean* based on the latest science, filling in the blanks based on his knowledge of modern fish and on instincts honed through a lifelong interest in dinosaurs. Troll was also breaking his own new ground, pushing paleoart into a realm of a self-styled, sometimes surreal expressionism by occasionally setting his creatures in oddball scenes that mashed past, present, and the dream state into one. Ruler-wielding nuns, dice, flaming cities, naked people, and cheeseburgers mixed it up with mosasaurs, pterosaurs, trilobites, and ammonites. While Troll thrived on artistic license, he also obsessed over scientific detail. It was a matter of pride that the creature he was illustrating be as accurately rendered as possible—even if it was floating over a school bus full of children, dancing with a barefoot girl, or playing guitar. For the *Helicoprion* that would appear in *Planet Ocean*, Troll was resolved to get it right.

J. D. Stewart hadn't been able to tell Troll much about *Helicoprion* —J.D. was a Cretaceous fish expert, and *Helicoprion* was a Permian chondrichthyan. But he suggested someone who could: a paleo shark expert named Rainer Zangerl, retired from the Field Museum in Chicago. In the early 1960s, zooming back to Chicago after a collecting trip, Zangerl had spotted an enticing roadside rock outcrop near the tiny town of Mecca, Indiana. Initial explorations yielded fossils, and Zangerl and a few colleagues found a landowner with a bulldozer to dig them a small quarry. They were astonished by what they discovered, and the aptly named small Midwestern town did indeed become hallowed ground for paleontologists. The Mecca Quarry Shale Member, as it's now called, was a rare *Lagerstätte*, a deposit of extraordinarily well-preserved fossils showing the mineralized remains of soft tissues like eyes, internal organs, skin, and muscle fibers. Buried in the Mecca shale layers was a dense concentration of previously unknown Paleozoic sharks, as well other fish and invertebrates. Zangerl had already established his professional reputation with fossil turtles, but the Mecca sharks eventually eclipsed his other activities. His work to excavate and describe the fossils led to his invitation to literally write the book on paleo sharks, as author of the *Handbook of Paleoichthyology, Volume 3A, Chondrichthyes I, Paleozoic Elasmobranchii*, published in 1981.

Soon after Troll returned to Alaska from his trip with Matsen, on a drizzly February day staring over the neighboring roofs out to the Tongass Narrows, he picked up the phone and dialed Zangerl hoping to learn more about *Helicoprion*.

"Yes, yes, I know J.D.," Zangerl said, his thick Swiss accent pushing through the coiled phone cord.

Zangerl was eighty-one. He had been born in Switzerland and received his PhD at age twenty-three from the University of Zurich. As part of his education, Zangerl learned how to use X-ray machines with the idea of using the technology in paleontological research. In 1937, he set off on a slow boat to New York, eventually making his way to Chicago, where, in 1945, he became the curator of fossil reptiles and amphibians at the Field Museum, then chair of the Geology Department from 1962 until his retirement in the mid-1970s.

One of the first burning, hopeful questions Troll had for this icon of paleo shark knowledge was if *Helicoprion* could lash out its tooth spiral. No, was the answer, absolutely not. Not to worry though. Even without the whipping whorl, the shark was a *fierrrce crrrr-itter!* Zangerl offered this happy assurance with the rolling *R*s of Germanic pronunciation. Undeterred, Troll explained his quest: to learn as much as he could about the whorl-toothed shark and create the most scientifically accurate *Helicoprion* illustration he could for *Planet Ocean*—a pursuit that would turn out to be as rife with pitfalls as trying to build a car from a broken steering wheel and one side mirror.

Zangerl was grandfatherly and friendly, with all the time in the world to talk about *Helicoprion* with a total stranger. The most unusual thing about this shark, the scientist said, was that it retained its teeth throughout its lifetime by reeling them up into the spiral. No other shark before or since was known to do that. *Zee oldest teese are at zee zenter of zee whorl*, Zangerl pointed out. It was one of the hardest ideas for Troll to digest because it felt so counterintuitive. Normally, smaller means younger. We're most familiar with growth cycles in which smaller forms get bigger over time. However, as biologists know, sharks' teeth emerge at their full size. But Troll was an artist, more in tune with how things look than how they work. Eventually, though, he got it.

Although no *Helicoprion* body fossils had ever been found, Zangerl advised Troll that he could reasonably base a *Helicoprion* body plan on sharks from the Caseodontidae family, for which body fossils did exist. Zangerl had some in his basement. The majority of large-bodied Paleozoic chondrichthyans, he noted, possessed the classic shark torpedo shape, with paired pectoral fins, prominent dorsal fin, and strong, vertical tail.

"You know," Zangerl said, "*Helicoprion* is a eugeneodontid shark."

No, Troll didn't know. In 1971, Zangerl had created a new order, Eugeneodontida, to corral all the sharks with curved, centered tooth structures that had been discovered since Joseph Leidy introduced the world to *Edestus vorax*. The Eugeneodontida order enfolded the Eugeneodontidae family and the *Eugeneodus* genus, named after Zangerl's friend and colleague Eugene S. Richardson.

Despite the plethora of "eugeneo" prefixes, when someone uses the term eugeneodontid they're usually referring to the whole kit-and-caboodle of sharks with curved, symphyseal dentition. Like the word "chondrichthyan," the term "eugeneodontid" can be useful around anyone who might object to calling *Helicoprion* a shark. But don't be lulled—because we're talking about paleo shark science, after all—into thinking that by using the term eugeneodontid you will necessarily be home free. Leading scholars working today think Zangerl's grouping is "artificial and potentially misleading," pointing out that there are two distinctly different kinds of "eugeneodontids," depending on how the teeth are locked together within the tooth whorl. *Nothing is settled!* they say.

Grabbing for ancient shark straws in the heady conversation, Troll mentioned megalodon (a mere sixteen million years old, but the only prehistoric shark Troll knew at that time) as the biggest shark of all time. *Don't be so sure*, Zangerl counseled. *Next to Edestus giganteus . . .* Troll scribbled this unknown name in his sketchbook, as a door in his mind blew open onto a new and previously unimagined ancient world of giant killer sharks where even the big-as-a-boxcar megalodon might be a small fry compared to *giganteus*. How come nobody knew about them outside a handful of super-specialized paleoicthyologists?

By the time Troll hung up the phone he was on fire for paleo sharks, especially the monster buzz saw beast that burst through the walls of his imagination when Zangerl told him—incorrectly, he would later learn—that *Helicoprion* had *two* whorls, one in the upper and one in the lower jaw. (It had been over a dozen years since Zangerl wrote the *Handbook*, and the *Helicoprion* entry was less than a hundred words, focused mainly on describing the teeth— one blip in a book stuffed with accounts of more than two hundred genera of ancient sharks.)

Troll got his hands on an interlibrary loan copy of Zangerl's *Handbook of Paleoichthyology*, which he opened like Pandora's box. Inside he found a reproduction of Bendix-Almgreen's mesmerizing *Helicoprion* tooth-spiral graphic, as well as reconstructions of parts and pieces from sharks in the Caseodontidae and Eugeneodontidae families—fragments of fin and tail and teeth and skull, like scattered belongings from a train wreck. There were bits of *Caseodus, Romerodus, Fadenia, Ornithoprion,* and *Sarcoprion*, and pictures of the bewildering *Edestus* blades. Also in the *Handbook* pages Troll discovered the Iniopterygia, an order of fantastical ratfish-shaped chondrichthyans about a foot long with huge heads, large eyes, paddle-shaped tails, large pectoral fins set high like wings, and barbs and barnacle-looking denticles scattered over their bodies. Zangerl and his colleague G. R. Case were the first to discover these "iniops," in 1973, in the Mecca shale.

The first *Helicoprion* Troll had drawn immediately after his conversation with Zangerl was a blunt-nosed shark with what looked like bicycle gears jammed into its nose and chin. Now the snouts on Troll's *helicoprion* grew longer and the whorls became more like circular saws. *Caseodus* had strong pectoral fins and a lunate tail with more-or-less equal lobes top and bottom, so *Helicoprion* got those too. Caseodontid fossils had been found with pronounced dorsal fins, but would the fin have been positioned somewhat forward on the back, over the pectoral fin like *Fadenia*, or slightly behind the pectoral, like *Caseodus*? As Troll made his choices, *Helicoprion*'s body emerged sleek and muscular, the whorls settled more organically into the jaws, and the animal gained a growing sense of motion and power. This was the rush. Here was the high. In a magical unfolding, Troll was the first person in the universe to see a long-lost creature from deep time. He was resurrecting the shark, right there in his sketchbook.

Troll drew furiously, not just *Helicoprion*, but a whole scene of Paleozoic chondrichthyans. In early March he sent a sheaf of drawings to Zangerl for feedback. Of *Helicoprion*, Zangerl's returned notes said: "Snout has to accommodate an entire spiral." "Too much exposed." "No." "No, no, no." Then finally a sketch showing the whorls well-imbedded in the jaws earned a "yes." As for the other sharks, Zangerl deemed Troll's *Sarcoprion* "excellent," and his *Promexyele peyeri* "quite good, but no rostrum." For *Iniopteryx rushlaui*, Zangerl shared that "The fish evidently carried a lure on its forehead." In his accompanying letter to Troll, Zangerl wrote that he was "particularly delighted to see some of the weird iniopterygians 'come to life.'"

Digging for more information on the ancient chondrichthyan bestiary, Troll discovered that there were no general-audience books on Paleozoic sharks. His research did, however, uncover references to Obruchev's 1953 paper and Bendix-Almgreen's 1966 paper, both of which he obtained through interlibrary loan. Obruchev's paper was in Russian but contained the figures illustrating the various ideas that had been floated as to where the whorl fit on the animal. Bendix-Almgreen's paper was in English—dense, mind-numbingly scientific English, but English—and Troll realized that according to the Dane's scrupulous research and high-quality fossils, *Helicoprion* had only one whorl.

Ja, okay, one whorl, Zangerl agreed, after he reread Bendix-Almgreen's paper. *Anyway, Bendix-Almgreen, pffft.* The two were feuding. Zangerl was aggravated at Bendix-Almgreen for disputing some of his ideas on certain taxonomic relationships among the Mecca sharks. "His arguments," Zangerl wrote in the *Handbook*, "can hardly be taken seriously . . ." For his part, Bendix-Almgreen was miffed at Zangerl for giving credit in the *Handbook* to Eigil Nielsen for drawings Bendix-Almgreen had done. Ah, the passions of science.

The *Planet Ocean* deadline was looming as Troll redrew *Helicoprion* with one whorl in the lower jaw and a small row of pointed teeth in the upper, and further refined his "Paleozoic Sharks" illustration, a crowded, colorful, time-scale-conflation of *Helicoprion* plus other odd chondrichthyans that the world would be laying eyes on for the first time—*Sarcoprion, Ornithoprion, Iniopteryx, Iniopera, Cobelodus, Romerodus, Orodus, Harpagofututor,* and more. Matsen had only given *Helicoprion* a paragraph or so in the book's final manuscript, so, determined to shine the world spotlight on his über beast, Troll wedged a last-minute whorl-toothed shark spread into the book, writing about finding the fossil in L.A., and illustrating a frowning Karpinsky with all the *Helicoprion* versions through time, from the original snout-whorl to Troll's New Year's party favor rendition.

Planet Ocean was published in 1994. Troll created a museum exhibit based on the book, which opened at Seattle's Burke Museum of Natural History and Culture, and featured the *Helicoprion* fossil whorl from L.A. County. Following its run at the Burke, the exhibit morphed into a traveling show called "Dancing to the Fossil Record." For the inaugural installation at the Denver Museum of Natural History (later renamed the Denver Museum of Nature and Science), Troll borrowed a fossil whorl from Idaho State University, laying a path that would circle back to great effect two decades in the future.

In the summer of 1994, learning that Zangerl's "dear frau" had died, Troll visited his first scientific mentor and now friend at his rural Indiana retreat, "Hajji Hollow," nearly on top of the Mecca Shale fossil beds. The name was a play on the honorific Muslim term "hajji," for someone who has completed the pilgrimage to Mecca. Zangerl was famous for his German-style dark bread, and a yeasty, fresh-bread fragrance met Troll when he arrived at the house. In the cinder-block basement, wooden shelves bowed under the weight of row upon row of black binders holding fossil X-rays. Zangerl picked up a slab of black shale the size of a lunch tray and showed Troll a *Caseodus* with preserved gut contents. At least that's what Zangerl said it was. Black-on-black shale fossils are notoriously inscrutable and unimpressive to the untrained eye, regardless of their significance. Even so, it was a hallowed moment. Troll and Zangerl walked the seventy-one-acre property splitting slabs of shale from Zangerl's own outcrops, a pilgrim and his teacher, looking for fossils.

———————

Where there had been none, the mid 1990s brought a small surge of general-audience books on fossil fish. There was *Planet Ocean* in 1994, then in 1995, Australian paleontologist John Long published *The Rise of Fishes,* an engaging homage to fossil fishes and their foundational role in the evolution of vertebrate life. In 1996, paleontologist John Maisey of the American Museum of Natural History published *Discovering Fossil Fishes*, another accessible and well-illustrated book about our aquatic ancestors and the progression of vertebrate design. Also in 1996, French paleontologist Philippe Janvier published *Early Vertebrates,* released as a textbook on the evolutionary relationships of early fishes but eminently readable. Long's and Janvier's books both included short mentions and head reconstructions of *Helicoprion*. Long illustrated his *Helicoprion* with an exposed tooth whorl curling under the shark's chin, and Janvier showed the whorl at the end of an elongated lower jaw, like *Sarcoprion*, with a set of curved symphyseal teeth in the upper jaw. Those were the days before Internet search engines, RSS feeds, online journals, and Web-based databases, and Long and Janvier must have missed Bendix-Almgreen's paper. Anyway, they weren't only thinking about *Helicoprion* for their books, but were considering five hundred million years of fish evolution, with its legions of shifting species and dazzling diversity, from the jawless, finless *Astraspis*, to the chondrichthyan forebearer *Cladoselache*, all the way through *Latimeria*, the only surviving genus of lobe-finned fish, the pioneering piscine class that led the way for us terrestrial vertebrates.

Troll wasn't a scientist, but through his focused passion, curiosity, and persistence, he had become *Helicoprion*'s most accurate portraitist, leading aficionado, and front man. In 1998, Troll and *Helicoprion* caught the attention of the Discovery Channel, which produced a short segment called "The Color of Ancient Sharks" for Shark Week. It was the tenth year of the program, but the first year with content on prehistoric sharks. In the segment, Troll stood in a vault-like collection room at the American Museum with John Maisey as Maisey pulled an *Edestus giganteus* fossil out of a box. The heavy fragment, a foot and a half long, bristled with six wicked-looking teeth. Next the cameras followed Troll to Montana to meet paleo shark experts Richard Lund and his research partner (and eventual wife) Eileen Grogan. Lund, then at Adelphi University in Garden City, New York, had spent decades digging a treasure trove of very weird, previously unknown Carboniferous chondrichthyans from a fabulous *Lagerstätte* known as Bear Gulch in middle-of-nowhere Montana. At the Bear Gulch field camp, Lund, Grogan, and Troll leaned together over a makeshift cable-spool table, where the pen-wielding Lund proceeded to cheerfully punch holes in Troll's *Helicoprion*. "The schnozz is going to come more like this." One other problem? "Get rid of these," Lund said, emphatically crossing out the gill slits on Troll's drawing. *Helicoprion*, said Lund and Grogan, in echoing assertion, has one opercular cover. One soft, opercular cover. Troll flinched. The operculum is the flap covering a fish's gills. Bony fish like snapper and salmon all have opercular covers, but the only modern chondrichthyans with opercula are the holocephalans, the ratfish.

A large proportion of the chondrichthyan fossils coming out of the extraordinary Montana quarry were in the holocephalan lineage. Lund and Grogan had coined a new name, "euchondrocephalan," to distinguish the huge and diverse group of ancestral (stem) holocephalans from the tiny group of modern (crown) holocephalans, which have been reduced to one surviving order, the Chimaeriformes, ratfish, with a meager fifty or so species. Lund and Grogan considered *Helico-prion* and the rest of the eugeneodontid group to be euchondro-cephalans. Although they hadn't found any eugeneodontids at Bear Gulch, all of their euchondrocephalan fossils—at least the ones that preserved this particular detail—had opercular covers instead of gill slits. Lund and Grogan were convinced that opercular covers should therefore be generalized to all euchondrocephalans. Writing about eugeneodontids in *Early Vertebrates*, Philippe Janvier had under-scored the diversity of the group, noting that, "Some were huge and sharklike, with a single prominent dorsal fin, whereas others were eel-shaped." Of the gill issue, he wrote, "At present, the number of gill slits remains unknown."

Troll's flinching angst around the spool table came from the fact that gill slits are visual shorthand for shark. Put gill slits on just about any fishlike form and people will identify it as a shark. Take the gill slits away and the shark gestalt—the predatory menace and pelagic power, the rhythmically flaring, breathing, beauty—evaporates. Lund and Grogan gave the demoralized Troll an eraser and sent him off to the campfire and back to the drawing board.

Troll tried. Back in his Ketchikan studio he created a five-foot by eleven-and-a-half-foot mural called *Swimming with Sharks*, in which the *Helicoprion* had an opercular cover—but it would be his first and last such depiction until he had more definitive news that slits were definitely out. While he was working on *Swimming with Sharks*, Troll was helping his young son learn the alphabet. The convergence sparked an idea for a shark alphabet book collecting the weirdest, quirkiest examples of living and extinct sharks, like goblin shark and *Helicoprion,* frilled shark, iniopterygian, *Listracan-thus,* megalodon, Queensland sawfish, and *Xenacanth*. Even in the now-crowded paleo shark pack, *Helicoprion* was still Troll's signature beast. He felt unsettled with his current *Helicoprion* illustration and he wanted to get it as close to perfect as possible for *Sharkabet*, so he reached out to the clear living authority, Svend Erik Bendix-Almgreen.

In summer of 2000, Troll mailed a set of *Helicoprion* drawings to Bendix-Almgreen, then sixty-eight years old and living in Copenhagen. "Dear Mr. Troll," the paleontologist wrote back, "*Helicoprion* is indeed fascinating and I can well understand its appeal to artists which like you try hard to re-create images of such creatures from long gone worlds. You ask for my comments regarding your attempted reconstructions of this particular late Palaeozoic fish and I shall note the following . . ."

The first thing Bendix-Almgreen nixed was Troll's depiction of sharp teeth in the upper jaw. Had such teeth existed, the eminent paleontologist noted, "it would undoubtedly have been preserved in position in the best of the specimens (Idaho no. 4) described and figured by me in 1966." Bendix-Almgreen had written his *Helicoprion* paper thirty-four years earlier, but he remembered the research and the fossils clearly. Idaho no. 4, he avowed, was so wonderfully preserved that if there had been well-developed upper teeth, it would have been obvious. There were blocks of flat little pavement teeth in the upper jaw, he said, but nothing else. Bendix-Almgreen advised Troll to settle the whorl more deeply into the lower jaw, and reiterated his original 1966 reconstruction of a slight extension at the end of the lower jaw, like the front fork on a bike, projecting slightly past the whorl.

"With respect to the remaining parts of your reconstructions showing the fish in its entirety, I can only regard this as pure conjecture," Bendix-Almgreen wrote. "You have apparently been inspired by some of Zangerl's figures (among which that of *Fadenia crenulata* is credited to my teacher, the late Dr. Eigil Nielsen, while actually it is mine! A fact Zangerl was informed about via Nielsen's letter!), but there is no evidence to show that the genus *Helicoprion* is a true close relative of forms like *Agassizodus, Fadenia, Sarcoprion, Erikodus,* etc. and certainly not of *Edestus* either."

Bendix-Almgreen was saying that just because Zangerl said *Helicoprion* was closely related to the caseodontids, it wasn't necessarily so. Then, as today, there is little scientific consensus on the relationships of Paleozoic chondrichthyans. But while Bendix-Almgreen could wait for more scientific evidence in his line of work, Troll, with a deadline for his *Helicoprion* illustration, could not. Without conjecture or inspiration, there would be no *H* for *Helicoprion*. Still, Troll wanted his postulations to be at least plausible, and he wanted Bendix-Almgreen's dispensation for his interpretation. Troll finagled Bendix-Almgreen's phone number from an administrative assistant at the Geological Museum of Denmark and caught the paleontologist and his wife at dinner when he placed the very long-distance call. *It's the artist from Alaska,* Bendix-Almgreen's wife said, and the cordial and accommodating expert left his supper to take the call.

Troll explained that he had gotten the detail of the upper teeth from Philippe Janvier's illustration in *Early Vertebrates*, and he fished for Bendix-Almgreen's approval in using the caseodontid body model. Bendix-Almgreen was polite, but unmoved. There were no upper teeth, and it was impossible to know what *Helicoprion*'s body looked like. No hard feelings.

For the *Helicoprion* in *Sharkabet*, Troll removed the upper row of sharp teeth and illustrated the front part of the lower jaw as Bendix-Almgreen had it in his paper, with tooth crowns protruding below the chin through the slotted extension. Troll continued to use the caseodontid body form as Zangerl had suggested, because he had to use something. And he restored the gill slits, because gill slits are a strong visual shorthand for shark, and he feared that depicting an opercular cover could aesthetically compromise his monster predator. *Sharkabet* was released in 2002, and at last American children had something besides dinosaurs to feed their prehistoric animal fascinations.

The previous decade had been electrifying and transformational for Troll, to listen to the scientists, add his artist's instincts, and will never-before-seen creatures to life under his pencil. Troll had visited, talked to, or corresponded with the full who's who of paleo sharks of the time—Zangerl, Bendix-Almgreen, Maisey, Janvier, Long, Lund, and Grogan—and he had collected, collated, archived, and artistically interpreted the existing paleo-shark body of knowledge. He was the whorl-tooth acolyte; the freewheeling monk zealously illuminating the Book of Lost Sharks.

In July 2004, Svend Erik Bendix-Almgreen died at age seventy-two. That December, Rainer Zangerl died at ninety-two. Feeling adrift as the lone, slightly lonely, keeper of the *Helicoprion* flame, Troll's attention wandered to other things. In the grand spiral of circumstance, the same year Troll's mentors died, a young marine from Mississippi returned to civilian life, stepping onto an unexpected path that would rekindle Troll's whorl-toothed flame. Meanwhile, on a different intersecting path, and in a fitting, full-circle sense, a Russian paleontologist kicked off a revamped de facto *Helicoprion* forum for the twenty-first century, with a beat-up fossil and fresh revelations. The new guard was assembling.

10

THE NEW GUARD

What of the sawfish? What on earth is that all about?
And the hammerhead shark?
—Richard Dawkins, *The Ancestor's Tale:*
A Pilgrimage to the Dawn of Evolution, 2004

IN 2009, THE GREAT NEEDLE-AND-THREAD OF CHANCE WHIP-STITCHED RUSSIA BACK into the *Helicoprion* story when paleontologist Oleg A. Lebedev published a paper on a previously undescribed tooth whorl. Lebedev stumbled across the fossil in a drawer at the Paleontological Institute of the Russian Academy of Sciences in Moscow. He had been a year old when paleontologist Vasiliy E. Ruzhentsev originally collected the specimen in Kazakhstan in 1958, apparently while mapping Permian-age rocks and hunting for fossil cephalopods. It wasn't the first or last time this paleontological coincidence between *Helicoprion* and cephalopods occurred. In 1995, Canadian paleontologist and ammonoid expert W. W. Nassichuk wrote, "Curiously, representatives of the coiled shark tooth structure *Helicoprion* occur with [Permian] ammonoids in both

Idaho and the Arctic Islands." Lebedev himself would soon reveal the conflux to be about as coincidental as fox tracks circling a chicken coop.

Ruzhentsev's fossil was a badly battered whorl broken into two sickle-shaped fragments. The larger fragment, a section of the whorl's outer perimeter, held evidence of twenty-six teeth, though all the crowns were broken off. The smaller piece was an inner section of whorl in noticeably better shape, with fifteen complete or nearly complete tooth crowns. Lebedev estimated that, in its intact state, the whorl would have been nearly sixteen inches in diameter with up to 140 teeth total. Ruzhentsev, not particularly interested in fish and possibly unimpressed by the much-deteriorated fossil, had given it to Dmitry Obruchev, a logical handoff since Obruchev had recently published his well-illustrated retrospective of Karpinsky's *Helicoprion* work. But Obruchev never published anything on Ruzhentsev's specimen, and the fossil lay unnoticed for nearly fifty years until it caught Lebedev's eye. While the fossil caught Lebedev's eye, the location of the find is what really captured his attention. Ruzhentsev found the fossil in the Kazakh region of the Ural Mountains, the same range but a significant distance away from the quarry where Karpinsky's stunning whorl had been discovered a century before. That made Ruzhentsev's fossil the southernmost *Helicoprion* ever collected on the western slope of the Urals, which indicated something very significant about *Helicoprion*'s distribution and range.

Here we recall the *Helicoprion* we met previously, Karpinsky's *Helicoprion*, swimming up the west coast of Pangaea. The big shark was following signals from its ampullae of Lorenzini, those jelly-filled sensory pores, to turn east into the Uralian Seaway—the thousands-miles-long "shortcut" from the Panthalassic Ocean to the Tethys Sea. According to Lebedev's estimation, Ruzhentsev found his *Helicoprion* fossil at what once had been the junction of the Uralian Seaway and the northern Tethys. Unlike the *Helicoprion* that ended up on Karpinsky's work table, which died about halfway down the seaway, Ruzhentsev's *Helicoprion* had navigated its entire length. Or perhaps was on its way back to the Panthalassic Ocean. Regardless of which way it was going, the discovery provided connect-the-dots evidence that *Helicoprion* was a "cosmopolitan" shark—global and widespread. Looking at a Permian globe, *Helicoprion* fossils had previously been clustered in three general areas: slightly north of the equator on Pangaea's west coast; near the west entrance, and in the western half, of the Uralian Seaway; and scattered around the Tethys Sea. Ruzhentsev's *Helicoprion* completed the circuit. In his paper, Lebedev plotted the world's *Helicoprion* fossil discoveries on a Permian map, and you can just about see the sharks swimming up the west coast of Pangaea past or over what is now Mexico, Texas, and Nevada, with some of them ducking into the Phosphoria Sea to die in what is now Idaho. Still more of the sharks pressed north, past what are now British Columbia and Alberta, Canada, swimming, swimming, swimming, turning east into the Uralian Seaway, with some of them making it all the way over to the Tethys to circle with the other sharks cruising the continental shelves of terranes that are now parts of China, Japan, Tibet, and Western Australia. Even though Ruzhentsev's *Helicoprion* wasn't nearly as complete or beautiful as Karpinsky's, it was nevertheless a dazzling find for what it revealed about *Helicoprion*'s geographic range.

An additional piece of the range puzzle serendipitously snapped into place in 2015, when a long-lost, off-the-radar *Helicoprion* from Alaska resurfaced. The plate-size fossil had originally been found in 1986, high on a scree slope north of the Arctic Circle, near Atigun Gorge, by twenty-three-year-old geology student and Iñupiaq native Richard "Savik" Glenn. Glenn and his adviser sent the fossil to the Smithsonian, where it was identified as *Helicoprion*, then mislabeled, put away, and forgotten—until thirty years later when Glenn, by then an executive at Arctic Slope Regional Corporation (as well as a whaling captain in Barrow), crossed paths with Troll in Alaska. The meeting sparked Glenn's memory of his *Helicoprion*, igniting an intense search during which a dogged Smithsonian collections manager opened drawer after drawer until he found it. The Alaska *Helicoprion* bridged the cluster of fossils on Pangaea's west coast with the cluster near the west entrance of the seaway.

The geographical distribution of *Helicoprion* hadn't even been a topic of discussion until the latter part of the twentieth century, because so little was known about the physical history of the earth until relatively recently. No one had pondered the shark's global citizenship. It was all about the location of the whorl on the shark, not the shark in the world. Meteorologist Alfred Wegener first proposed a theory of continental drift in 1912, and over time, other geologists more correctly understood and refined his ideas into the model of plate tectonics, defined in a set of papers written between 1965 and 1967. The explanatory and predictive power of plate tectonics was groundbreaking, and revolutionized the earth sciences. In a nutshell, the earth's rind (the lithosphere, comprised of the crust plus a thin rigid skin of mantle) is broken into six major and numerous smaller tectonic plates. These plates float and jostle on convection currents of molten rock that flow with a consistency of liquid asphalt. Most of the large plates carry a combination of land and ocean. For example, the North American plate is like a

Thanksgiving platter loaded with scoops of food awash in gravy. Our home plate, which nudges west-southwest about 2.3 centimeters a year, carries North America, Greenland, and the northern Caribbean, as well as much of the North Atlantic and parts of Siberia, Iceland, the Azores, and the Arctic Ocean. Over the history of the planet, the plates have docked in various configurations, including Gondwana, Pangaea, and the arrangement we know today. Tectonic forces uplift mountains, give birth to volcanoes—and closed the Uralian Seaway when Pangaea began to rift apart at the end of the Permian. It was no loss to the *helicoprions*, since they were extinct by then anyway.

In his paper, after establishing the whorl-toothed shark as a global citizen, Lebedev took up the question of where *Helicoprion* fit into the picture ecologically. Basically, the question of ecological niche is a kind of chicken-or-egg question of diet. What an animal eats determines where and how it lives. Or where and how it lives determines what it eats. For *Helicoprion*, the more interesting chicken-or-egg question wasn't what it ate, but *how* it ate. How did it actually use that crazy whorl?

Karpinsky had initially visualized the whorl as a weapon to attack and fight other fish. C. R. Eastman's 1902 counterargument that the whorl had a "chiefly masticatory" function broadly replaced Karpinsky's proposition, but details weren't forthcoming. Obruchev and his next-generation Russian counterpart B. I. Tchuvashov, who in 2001 described the poorly known eugeneodontid *Shaktauites seywi*, thought *Helicoprion* might use its whorl to plow invertebrates out of the substrate. It's an awkward move to imagine, since both Obruchev and Tchuvashov pictured the whorl curling over the upper snout as Karpinsky had, but the feeding behavior of the "living fossil" coelacanth fish includes a headstand feeding position, so why not? Bendix-Almgren, for his part, wrote that the whorl seemed built for cutting and tearing soft prey, but he didn't expand on the idea. Australian paleontologist John Long compared whorl function to the rostrum in the very strange, still surviving but very much endangered sawfish, which slashes through schools of prey fish to stun and kill them. *Helicoprion* may have used its whorl, Long wrote, "to thrash about and snag squid-like creatures or other fish on the jagged teeth."

To come to his own conclusions, Lebedev conducted a microscopic study of the best-preserved tooth crown surfaces on one of Karpinsky's original fossils. Lebedev didn't find marks—either of the type, or oriented in a direction—that would show evidence of plowing, but he did see "scratch traces" aligned in a way that indicated bite force was applied between the jaws. "Thus," he wrote, "the whorl might be used for grasping prey or tearing off large pieces by pressing the food object against the upper jaw. Serration of the crown edges facilitated cutting soft tissues." *Helicoprion,* Lebedev suggested more specifically, was largely eating soft-bodied cephalopods like squid, or the exposed head and arms of shelled cephalopods like ammonoids. Those faint scratches on the tooth crowns might have resulted from incidental contact with shells.

Lebedev found his "morphonutritiological" model in the odontocetans, which are toothed whales like dolphins and orcas. Going straight to the top, he presented as evidence the sperm whale—the world's largest toothed predator, at up to fifty-two feet long. The sperm whale has teeth only in its very narrow lower jaw, which clearly is a highly effective arrangement for catching and consuming its most common food, giant squid and other cephalopods. Odontocetan biologists have observed that the fewer the teeth, the greater the percentage of cephalopods in the diet. "It is therefore not impossible," Lebedev wrote, that *Helicoprion* and its kin occupied the same ecological niche in the Permian as whales do today—squid hunters. (Whales didn't evolve to claim the niche until the Cenozoic era, another two hundred million years into the future.)

Lebedev capped his paper with his own full-body reconstruction, the first by a scientist. He based the body, as Troll had, on the three eugeneodontids known by more or less complete skeletons, *Caseodus, Romerodus*, and *Fadenia*, from Zangerl's *Handbook*. Lebedev's *Helicoprion* is a stately, grandfather-looking shark, with gill slits and a hawkish snout. Also at the end of his paper Lebedev offered a formula for determining the shark's size, based on the relationship of whorl diameter to lower jaw in those above-named eugeneodontids, and working under the assumption that *Helicoprion* had sharklike proportions, and was not elongated like an eel. The largest whorl diameter he knew of was a little over a foot, which would correspond to a skull length of three to five feet, to predict an overall body length ranging from sixteen to just over twenty-six feet. That would have been impressive enough, but Lebedev's largest whorl wasn't the half of it.

––––––––––

It was October, 2010. Jesse Pruitt was in his third semester as an undergraduate at Idaho State University, working as an intern at the university's Museum of Natural History. This was his fresh start following his stint in the marines, and after his shoulder injury. For his first foray into undergraduate research, he had just gotten hooked on unraveling the riddle of a big, rough-looking slab of a fossil holding a spiral of giant teeth.

Pruitt didn't know it, but that hundred-pound hunk of rock, IMNH 36701, was one of the world's largest *Helicoprion* fossils, with a whorl close to two and a half feet in diameter, coiled in four and a quarter volutions. By comparison, Karpinsky's famous Russian whorl had three complete volutions and was only about ten inches in diameter. The outermost teeth were wickedly big: an inch and a half wide at the base and over four inches tall—longer than a playing card. Some 270 million years earlier, on what was probably an otherwise unremarkable day, 36701 had turned from the open Panthalassic Ocean into the shallower waters of the Phosphoria Sea. Maybe it was chasing a *Uraloceras nevadenense*—a fast-swimming ammonite of the time. Maybe it was a pregnant female looking for calm nursery waters. We'll never know why the big fish swam into the Phosphoria, but we know it never swam out. Ages later, its tooth whorl was unearthed at the Gay Mine on the Fort Hall Indian Reservation northwest of Pocatello, Idaho. Operating from 1946 until 1993 under leases between the Shoshone-Bannock Tribes and the J.R. Simplot Company, the mine, named after Simplot's daughter, was the longest operating open-pit phosphate mine in Idaho, processing almost twenty-five hundred acres over its lifetime, and passing along fourteen *Helicoprion* whorls to the museum.

The massive 36701 was cracked and shedding flakes of brown rock and darker brown fossil. Remarkably, it even held mineralized remnants of the shark's tooth crowns and root-base. Research begins with questions, and Pruitt had plenty. First up, did the whorl form in a taphonomic process after the shark died? Taphonomy is the study of what happens to an organism between its death and fossilization. As a close observer of nature, as well as a hunter and fisherman, Pruitt had seen firsthand how organic forms shifted after death, a transformation that often involved curling. His first theory was that *Helicoprion*'s teeth might have projected out in front of its body, like the studded rostrum of a sawfish, and that the appendage curled into a spiral after the animal died. To test his idea, he thought he would try to obtain fresh shark cartilage that he could expose to different environments to see what happened. Tied to the idea of the straight, tooth-bearing rostrum, Pruitt speculated that the smallest teeth, those at the center of the whorl, were the youngest, and that they grew larger and moved forward on the presumably straight rostrum as the shark aged to eventually shed off the front. Pruitt found an analog in elephants, whose teeth emerge at the back of the jaw and migrate to the front to fall out.

Fresh shark cartilage and elephants were in short supply in Idaho, and Pruitt was asking questions no one could answer, so Thompson suggested he contact Ray Troll. Thompson had become aware of Troll a decade earlier when the museum loaned him a whorl for the "Dancing to the Fossil Record" exhibit, and people at the museum remembered Troll as the *Helicoprion* headmaster. Pruitt found Troll's email address through his website and sent him a note explaining that he was doing a research project on *Helicoprion* and outlining some of his questions.

It's a long story, Troll replied. *Give me a call.*

On their first phone call, Pruitt thought Troll was a little weird and a bit gruff. He could hear Troll cringing when he laid out his initial ideas—that in life *Helicoprion* had a straight bill that curled into a whorl after its death, and that the smallest teeth must be the newest teeth. No, said Troll. Oh, no. No doubt recalling his own struggle with that concept, he channeled Zangerl into the phone: *Zee oldest tees are at zee zenter of zee whorrrl!*

Pruitt seemed to be listening carefully and Troll took note that he was at the Idaho Museum of Natural History, which had helped him out nearly a decade before with a fossil for his show, so the artist told the aspiring student that if he was truly serious, Troll would be happy to tell him all the ways he was wrong. Like King Eurystheus handing Hercules the Twelve Labors, Troll gave Pruitt a list of academic papers to track down and read, and told Pruitt, a green undergraduate intern in Idaho, to talk to the leading lights of the paleo shark world: John Maisey of the American Museum of Natural History, Australian paleontologist John A. Long of Flinders University, and Michael Coates of the University of Chicago.

Pruitt rose to the challenge. His outreach to the scientific lions brought mixed results. Maisey suggested Pruitt was probably wasting Pruitt's own time as well as his by asking about *Helicoprion*; the current knowledge was in a settled state. Long more cheerfully agreed that until new fossils were found there wasn't anything more to discover about the beast, but good luck. (In 1987, Long and Maisey had ventured into the Australian outback looking for *helicoprions* themselves, without success.) Most helpfully, Coates offered insights into the study of Paleozoic sharks in general, and the family of sharks that included *Helicoprion* in particular. Some of the early sharks left detailed fossil records, Coates explained, even back to the four- to six-foot-long Devonian predator *Cladoselache*, whose fossils preserved body shape, muscle fibers, stomach contents, and the all-important cranium. (All-important because the cranium reveals what evolutionary path the animal followed up through the ranks of class Chondrichthyes.) In contrast, the whorl-toothed shark bodies didn't seem to fossilize well, leaving nothing for paleontologists to puzzle over besides the tooth spirals. Coates sent Pruitt a number of publications about other Permian sharks, with the caveat that in order to make any real progress in understanding *Helicoprion*, someone would have to come up with something other than teeth.

But teeth were what Pruitt had, plus his own long list of questions to sort out. Like, did the whorls grow as paired structures? It wasn't such a wild idea—C. R. Eastman had asked the same question a century earlier. Pruitt began to look for evidence in the whorls that might indicate they came in pairs, soon turning to Leif Tapanila for help when his own tricks and ideas ran out.

At the time, Tapanila was the museum's research curator and associate professor of geosciences. Born and raised in Ontario, Canada, the thirty-four-year-old Tapanila had earned his PhD from the University of Utah five years earlier, focusing on trace fossils, which are the preserved burrows, diggings, tracks, and trackways left behind by long-gone creatures. Tapanila was one of about two people in the world specializing in Cretaceous clams and snails—putting him in a unique position to help inform the far larger pool of dinosaur specialists about what sorts of environments their beasts were running around in. He had loved math as a boy, and was fixed on being a mathematician when he grew up. The vision included teaching, since both his parents were teachers. A poor instructor in eleventh grade drove him into a biology class, however, where he discovered a scientific expression for his love of animals. The next twist in his path came as an undergraduate at the University of Waterloo in southern Ontario, when he took a geology class and it clicked for him—he could study animals across time. Tapanila had seen a picture of a *Helicoprion* whorl in a textbook on vertebrate paleontology, but the teacher glossed over it, saying something like "Paleozoic sharks are weird," before the class raised its collective eyebrows and moved on.

The next time Tapanila saw a *Helicoprion* whorl was in a small display at the Idaho Museum of Natural History when he arrived at Idaho State in 2005 as a visiting assistant professor. The curator at the time mentioned they had a bunch downstairs, and that was the extent of the conversation. Engrossed in his own projects, Tapanila assumed that all the work that could be done on the whorls had been done, and hadn't thought any more about the fossil whorls until Pruitt came along asking questions. Good questions.

While Tapanila didn't have any specific answers on *Helicoprion*, he was happy to help Pruitt shape and execute the research project. When Pruitt described his idea about paired whorls, Tapanila introduced him to bilateral symmetry, the body plan common to 99 percent of animals. (John Strong Newberry et al. had raised the topic of bilateral symmetry during the *Edestus* "fin-spine-versus-tooth-structure" debates.) Bilaterally symmetrical animals can be divided vertically into approximate mirror-image halves, with left and right tusks, ears, eyebrows, feet, pectoral fins, ribs, and so on. (Although the vast majority of animals are bilaterally symmetrical, the adult forms of sea stars, jellyfish, and the like are "radially" symmetrical—that is, their symmetrical parts radiate around a central axis, like slices of a pie or sections of an orange. Sponges are the only animals with asymmetrical body plans.) Tapanila explained that if *Helicoprion*'s whorl was a paired structure, the spiral would show "translation"—a slight bulging on one side that would have occurred as the whorl grew, to shape it into a right-side or left-side feature. If the whorl was a midline structure, on the other hand, it would grow in a flat plane, like a discus.

Tapanila helped Pruitt find research papers looking into related morphological questions so he could see how others had approached similar investigations, and the two worked out a method to measure whorls for translation. Pruitt dug in, working on his own but meeting regularly with Tapanila to report his findings and brainstorm. It soon became clear there was no translation, but the two were working well together and agreed to push on with data collection. They decided to create a protocol for measuring various elements of the whorls in the collection and see what might come out of it. To find out what sort of sample was on hand to measure, Pruitt and Tapanila went down into the basement of the museum. Tapanila punched the code into the keypad lock, and they walked through the fossil preparation rooms where Pruitt had been working, through another door into the humidity-controlled collection room, making their way to a bank of deep metal shelves in the far back corner. This is where the oversize specimens were stored, not sorted by geologic time, unlike the smaller specimens that could be cached in drawers or on shallower shelves. The nearest overhead bank of fluorescent lights was closer to the center of the room, so it was a little dim, especially in the shadows of the bottom shelf. On their hands and knees they peered in at enormous caramel-colored herbivore jaws, heavy hunks of rock with elegantly curled ammonites, and dark stone slabs with scale impressions from giant fish.

Only five years apart in age, Pruitt and Tapanila could hardly have been more different on the surface. Tapanila, the intense, whip-thin Canadian, stayed on the academic path from high school to PhD, and had already written or coauthored more than twenty peer-reviewed papers. Pruitt, the soft-spoken, laid-back Mississippi boy built like a fullback, joined the marines right out of high school and was in his third semester of college, at the age of twenty-nine. But inside, they were much more alike—intellectually curious, collaborative, hard-working, and open-minded. With their complementary skill sets and experience, they had a lot to offer each other. Tapanila sat back on his heels as Pruitt pushed aside the camel jaws and ammonites to strong-arm the *helicoprions* out where they could see them—one, after another, after another. Between the large whorls they dragged off the bottom shelf and the smaller fossils they found in drawers, there were about thirty specimens. It was the largest collection of *Helicoprion* fossils in the world—even before acquisition of a private collection in 2013 would bump the number to seventy-five.

Realizing the wealth of material they had, Tapanila and Pruitt submitted a proposal to ISU and won an undergraduate research grant to study the fossils in detail. The project would involve some chemical analysis and electron microscopy, but the core objective was to create a *Helicoprion* database—a unique set of measurements recording a broad data suite across multiple whorls. They would use geometric analysis, a recently developed mathematical tool for analyzing curved surfaces that measured shape using landmark coordinates rather than linear measurements. Establishing their protocols and terminology, Pruitt began measuring, measuring, measuring, with digital and traditional calipers. For days on end, Pruitt was on the floor with the big fossils or hunched over a table with the smaller ones, diligently measuring and making notes. Pruitt would plot initial graphs and share the numbers with Tapanila, who would analyze them further, walking Pruitt through the more complicated math. They started to see some interesting patterns and distinctions among the whorls, and began to think they might be able to make a genuine contribution to science with the work. Ten separate species had been named since the first *Helicoprion* discovery, but there were no clearly defined parameters for species assignment. So Tapanila helped Pruitt narrow his research focus to specifically target that question: What distinct characteristics set *Helicoprion* species apart from one another?

Tapanila made sure the work was rigorously done. He was an invertebrate specialist—an "invert guy," as they call themselves—and Pruitt was an undergraduate, yet here they were pushing into a frontier territory of vertebrate paleontology. They were maverick outsiders, so everything had to be scrupulously executed. They measured, graphed, and analyzed about two dozen specimens, capturing such data points as tooth height, width, and angle of blade at the tip. They compared large whorls to small whorls, counted teeth on the intact whorls, and looked at the overall whorl shape and size, including volution height, as determined by logarithmic angle. One of their more interesting initial discoveries was that, across all the specimens, teeth on any given whorl were the same shape until tooth number eighty-five. After that point, individual tooth size, shape, and proportion changed progressively with each successive tooth. Each new tooth crown was bigger than the one that preceded it. The height difference between tooth number one hundred and tooth one hundred and two might be nearly three-sixteenths of an inch. Pruitt and Tapanila concluded that tooth eighty-five must mark the puberty line where the shark shifted from subadult to adult. It seemed there might be more to learn than they had first thought.

Pruitt had been staying in touch with Troll while the research was under way, talking sharks and bouncing ideas and information back and forth. Troll urged Pruitt and Tapanila on, peppering them with questions and advice. Tapanila didn't know much about Troll, and was initially uncertain about his involvement. Clearly he was passionate about *Helicoprion*. But what was his role here? What was his angle? Soon enough Tapanila understood that, despite the fact that he wasn't a scientist, Troll seemed to hold the keys to the *Helicoprion* kingdom, thanks to his years of dialogue with the past generation of experts. He knew the research, and he knew all the players in the paleo shark world. It was an unconventional alliance but energizing and constructive.

Late in the winter of 2010, in a phone conversation between Troll and Pruitt, when Pruitt seemed stymied by the shark mystery, Troll put on the full-court press. *Have you read Bendix-Almgreen's paper yet? Read Bendix-Almgreen's 1966 paper! He had it figured out. Look at Idaho no. 4. You have it right there. The head is in there, man!*

Michael Coates's admonition came ricocheting back into Pruitt's mind: if you want to move the science forward, you have to find something other than teeth. In this case, "find" meant ransacking for Bendix-Almgreen's paper, grabbing Tapanila, and running down to the basement. If the fossil really did preserve jaw material as Bendix-Almgreen said it did, they might be able to finally pin down where and how the whorl fit in *Helicoprion*'s mouth, and maybe even how it worked. It took them a suspenseful few minutes to put their hands on it—the pictures in Bendix-Almgreen's paper didn't exactly match the specimen because some additional fossil preparation had been done. But after more slowly sorting through the rocks again, there it was, just as Troll had said. Their significant new piece for the puzzle: Idaho no. 4.

II

RESURRECTION, ONE SLICE AT A TIME

I was never particularly struck by dinosaurs.
They always seemed to me a bit modern, not mysterious enough.
So the further back you could go the better, for me.
—Jenny Clack, paleontologist, 2002

TAPANILA AND PRUITT HOVERED OVER IDAHO NO. 4, THE NEWLY BEAUTIFUL HUNK OF gray rock holding the final traces of a shark laid down in death long, long ago. They peered at the textures, ran their fingers over the surfaces, and scrutinized the rock's broken edge—seeing the fossil with new eyes, as if for the first time.

Opening Bendix-Almgreen's paper like a treasure map they thumbed to the back, to Plate I, where dashed lines pointed to different places in the photo of Idaho no. 4 marking gold: *calc. cart. l.l.j.* Calcified cartilage, lower left jaw. *Calc. cart. l.r.j.* Calcified cartilage, lower right jaw. Some half-dozen dashed lines arrowed-in around the circumference of the fossil, indicating what Bendix-Almgreen had interpreted as being preserved cartilage. Tapanila and Pruitt could clearly see those areas on the fossil in front of them, swatches of finely pebbled, uniform texture, different from the surrounding rock.

This was what Michael Coates had talked about. What John Maisey and John Long said they needed. It wasn't a new fossil. It wasn't even a new observation. But it was a certifiably exciting discovery—a fortuitous convergence, a wormhole of opportunity shimmering open at this place, in this time, for these paleo-practitioners. Bendix-Almgreen had written about it, now Tapanila and Pruitt were looking at it: promising indications—if Bendix-Almgreen's notes were correct and those textured areas and lines really represented fossilized cartilage—that at least parts of the upper and lower jaw, and perhaps even front parts of the cranium, were preserved in Idaho no. 4. They just might have the *Helicoprion* Holy Grail in their hands.

Troll was ecstatic when he heard they located the fossil. He could feel Zangerl and Bendix-Almgreen stirring and felt his own passion for the shark, quiescent for the last handful of years, thump alive in his heart and imagination. He all but booked a ballroom for the revival of the *Helicoprion* fan club. *Scan that sucker!* he urged Tapanila and Pruitt. *With what money?* they replied.

Paleontologists had been utilizing high-energy X-ray computed tomography (CT) to tease information out of solid rock since the late 1990s, when paleontologist Tim Rowe began concerted efforts to use CT technology like Superman vision on fossils. In CT scanning, multiple X-rays are taken from different angles then assembled into a cross-sectional image. The technology was originally developed for medical applications to examine soft tissue and bone. With CT, doctors could scrutinize virtual slivers of the human body using X-rays to make their slices instead of scalpels. (*Tomos* is Greek for slice, or section.) The first scan was of a brain, in 1971 in Wimbledon, England.

In short order the technology was adapted to industrial uses, typically for detecting flaws, analyzing failed parts, and reverse engineering. Freed from concerns over radiation levels and squirming patients, engineers were able to increase and more narrowly focus X-rays in longer exposures. As a result, today's high-energy, high-resolution scans capture details smaller than a human hair, even in high-density materials—like the phosphate-heavy rock that held Idaho no. 4. Advances in industrial CT technology opened the door to paleontological applications, and in 1997, Rowe and William Carlson cofounded the pioneering High-Resolution X-ray Computed Tomography Facility (UTCT) on the University of Texas campus in Austin. The first year UTCT was open for business, four papers were published incorporating data gleaned from UTCT scans, all authored or coauthored by Rowe or Carlson. Now as many as seventy papers a year are published in peer-reviewed journals by scientists from around the world who have used the UTCT facilities to uncover new information about their subjects of study—like "The endocranial morphology of the Plio-Pleistocene bone-cracking hyena *Pliocrocuta perrieri*: behavioral implications."

CT scanning is expensive, however, and was clearly beyond the budget of Pruitt's undergraduate research grant. Scanning for academic projects at UTCT cost $120 per hour, and one scan—of a large *Helicoprion* whorl in dense rock, for example—could take forty to sixty hours. And scans are just the first step, the first expense, in the process. That bulging grab bag of X-ray slices has to be reassembled in what has been described as a sort of Rubik's Cube process, which costs extra.

Troll kicked into producer gear. This was personal for him. He had been carrying a torch for *Helicoprion* for almost twenty years, now here was a chance to get "his" shark some twenty-first century scientific bona fides, as well as the respect, attention, and love he felt it deserved. Dinosaur-grade attention. He would find the money.

In March 2011, Troll traveled to Pennsylvania to install the "Cruisin' the Fossil Freeway" exhibit, based on the book he did with Kirk Johnson, then chief curator at the Denver Museum of Nature and Science. The exhibit would be at the North Museum of Nature and Science in Lancaster, Pennsylvania, and Troll's old friend Dominique Didier happened to be teaching biology and ichthyology at nearby Millersville University. Troll offered to give a "visiting fish artist" presentation to Didier's students while he was in the area, and Didier offered him a place to stay. Didier was one of the world's foremost—one of the world's only—authorities on modern chimaeras. Even before they met in the 1980s, Didier had appreciated Troll's ratfish art, and Troll had read Didier's ratfish studies. One day a large package arrived at Didier's door with a ratfish sketched on it, and inside was a lab coat bearing a silk-screened image of Troll's latest ratfish art. She wrote him back and they had been friends ever since. Troll affectionately called her the "Ratfish Queen."

Culturally, we have built an image of scientists as a tribe of pale lab dwellers sprung from serious, bookish children. But the vivacious, rollerblading Didier, like legions of earth scientists before her, began life as a wild child, running around outside playing in the dirt. For Didier, that dirt was in a small town outside Chicago. Her science career was launched when her favorite aunt, a Catholic nun and elementary school teacher, dropped off a bunch of jars saying she needed specimens for her classroom. Whenever Didier found anything interesting that summer—an insect, worm, or odd something-or-other—it went into a jar for Sister Ambrose. A few years later, after learning about muscle construction in high school biology, Didier built a model of myofibrils (rodlike filaments found in muscle cells) out of toothpicks and tin cans. Another handful of years later she was at a microscope at the Academy of Natural Sciences, an associate curator, squinting at the muscle fiber in ratfish embryos.

As an undergraduate at Illinois Wesleyan, Didier wanted to study bats. Her adviser, himself a bat expert, steered her into the wide-open field of chimaeras instead, because too many people were already studying bats. No one was looking at chimaeras, he said. *No wonder,* she thought. She found them disgusting, with their huge reflective eyes, gopher-shaped heads, rabbitish mouths, ratlike tails, and shimmering, almost translucent skin. They deserved their name, derived from the fearsome Greek monster Chimaera, cobbled of various creatures. She hated them. But the professor was insistent, and right. A few ratfish papers had been published in the 1960s, but the most thorough research on chimaeras was still Bashford Dean's classic 1906 book, *Chimaeroid Fishes and their Development.*

"Every investigator will admit that Chimaeroids have been but little studied—surprisingly little studied," Dean wrote, as if speaking to Didier from the shadowy past. Not only did Dean study ratfish, he had an "ardor" for them. And so, ultimately, would Didier. As she dove into the work it became like an arranged marriage that turns into a passionate affair. Fast-forward through a PhD in zoology from the University of Massachusetts at Amherst, and Didier not only had come to love ratfish, but she also had taken her place as the world's leading expert on chimaera diversity and evolution. She had described seventeen species of chimeroids, even naming one ratfish for Troll, *Hydrolagus trolli,* in recognition of his "valiant efforts to increase ratfish awareness worldwide." Found a mile deep in the Southern Ocean, the highly distinctive "pointynose blue chimaera," as it's commonly known, could reach lengths of more than three and a half feet. Didier also named a chimaera in honor of the professor who directed her away from bats and toward ratfish. That one, an almost black fish found off the coast of Australia, Didier designated as *Hydrolagus homonycteris—homo*, man, *nycteris*, bat. Bat man.

On the day of Troll's visit to her class in Millersville, Didier expected a talk about ratfish, but Troll gave a jazzy presentation on *Helicoprion* instead. The next morning, at breakfast with Didier and her then boyfriend (now husband), Troll couldn't stop talking about *Helicoprion* and the work this unlikely undergraduate student, Jesse Pruitt, and this invert guy, Leif Tapanila, were doing in where-the-hell Pocatello, Idaho. Troll told Didier about Idaho no. 4 and the jaw they thought was buried in the rock—the jaw that could unlock the *Helicoprion* story if only they could find the money to CT scan it. Hint, hint. At the time of Troll's visit, Didier happened to be one of five principal investigators working under a National Science Foundation grant to sort out chondrichthyan lineage for the Tree of Life program. The ambitious NSF "megascience initiative" had set no less of a goal than mapping the evolutionary relationships of all 1.75 million species of living and extinct organisms ever discovered on the planet. The argument was that a fuller understanding of the evolutionary relationships of all life-forms would benefit not only theoretical science, but also could help identify and fight disease, improve global agriculture, protect ecosystems from invasive species, aid in species conservation and ecosystem restoration, and even develop more effective antivenins for snake bites.

Even though Didier studied chimaera evolution, she was not a fossil-loving "rock jock." She was far more interested in living ratfish than dead Paleozoic sharks. But the whorl images and *Helicoprion* art Troll shared with her class were beguiling, and Troll was so fervent, it was impossible not to feel inspired and on the brink of something exciting. Didier told Troll to have Pruitt and Tapanila contact her to further explain what they thought they had, and what they hoped to learn. That they did, and the more they talked, the more she felt motivated to help. Hard evidence is extremely scarce in the realm of ancient sharks, and Didier realized that this unlikely pair in Idaho might be sitting on important new evidence that could advance the understanding of one of the world's most unusual and poorly known groups of Paleozoic chondrichthyans. CT scans of Idaho no. 4 could shed light on evolutionary relationships of early chondrichthyans, and as such, it might be possible to swing some funding from the NSF grant for Pruitt to take the fossil to UTCT in Austin.

The Idaho State campus did house a scanning laboratory, the Idaho Virtualization Lab, but IVL did surface laser scans, not internal X-ray CT scans. Laser scanning was great for modeling objects like Pleistocene horse craniums, grizzly bear bones, arrowheads, and woven baskets. IVL's online comparative osteological database, which allowed researchers to compare different kinds of bones, was the first of its kind in the world and included the first digitized and articulated orca and humpback whale skeletons. The bone database supported IVL's mission to "democratize" science by making cultural and biological artifacts digitally available to anyone with an interest, at no cost, with no university affiliation required. Unfortunately, laser scanning, however cool it was, wouldn't crack the code on Idaho no. 4.

Coincidentally, IVL manager Robert Schlader had been experimenting with combining external laser scan and CT data to generate 3-D computer models, particularly of skulls. Schlader had CT processing software at IVL, and to get scans, he was taking objects up the road to Bingham Memorial Hospital in Blackfoot, where technicians scanned them after-hours. One night Schlader and Pruitt took a few *Helicoprion* fossils to the hospital for scanning to see what they might get. The scans picked up material imbedded in the rock, but the resolution of the medical scanner wasn't high enough to capture clear details. Pruitt shared those scans with Didier as a sort of proof of concept—there was *something* in there—but Didier agreed the information was too thin to be useful. They definitely needed better scans.

Didier managed to shake loose some money, and one morning in May 2011, Pruitt put Idaho no. 4 and two other *Helicoprion* fossil whorls on the floor of the backseat of his little Dodge sedan and set off for Texas. His intention was to make the 1,450-mile drive in one shot—because he was young and because there was no money for hotels coming or going—but fatigue overtook him in New Mexico. He pulled into a rest area outside Roswell, UFO capital of America, and fell asleep in the car hoping this wasn't going to be a wasted journey. The UTCT scan tech had warned them not to expect too much because of the density of the phosphatic concretion in which the *Helicoprion* fossil was imbedded. As dawn rose over New Mexico, Pruitt stretched, pulled out of the rest area, and detoured into the tiny town of Roswell to check it out. Behind him in the car sat three mysterious emissaries from Earth's inner *and* outer space—rock layers composed of elements tracing back to the very formation of the planet from the universe's abundant supply of gas and dust. Out the window he saw a McDonald's shaped like a flying saucer, and a fake UFO tower in the town's center. Unimpressed, Pruitt drove on to Texas.

On Monday morning he delivered the fossils to the UTCT scan tech, who explained the CT scanning process. In turn, the tech listened as Pruitt covered safe-handling procedures for the fossil, which was pretty much just don't drop it. The scanner was like a lead-shielded room within a room, with machines and gears, servos and cables, and just enough room to walk in, turn, and set the fossil on a special table. Basically it was a thick-walled, twelve-foot by twenty-foot X-ray machine that could accommodate specimens up to the size of a bas-ketball. Most of the day was consumed getting things up and running, stabilizing the heavy specimen on the table, and running test scans. There was initial trouble getting the software tuned in—it turned out the emitter and detector on the scanner were slightly out of whack, and due to be recalibrated the following week. Pruitt ground his teeth over that news, and stepped outside to blow off steam on a phone call with Tapanila. If he had known, he could have easily scheduled his trip for after the machine was realigned. Happily, the tech was able to synch the machine well enough, and when he ran a final test through the middle of the rock, Pruitt forgot his frustration as beautiful light gray lines appeared running through the center of the fossil.

That evening Pruitt ate some Texas barbecue, watched a million or so bats—the largest urban colony in the world—fly out from under the Congress Avenue Bridge, and slept in a bed in a motel. Tuesday morning he worked with the scan tech to reorient the fossil to maxi-mize the potential for capturing data, then they let the machine run its task of X-raying 360 slices, at about eight minutes per slice. Each slice had three views, based on three scanning axes: front to back, top to bottom, and left to right. When it was finished two days later, Pruitt retrieved Idaho no. 4 and one of the other whorls. He left the biggest fossil with the tech, who was going to try to figure out a way to scan the oversize specimen, which also appeared to preserve cartilage. With the scan data in hand on a Jumpdrive, Pruitt headed for home.

A pile of CT scan slices doesn't mean anything until you put them back together. At that point, still hoping to have UTCT process the scans and create a 3-D computer model, Pruitt and Schlader dug in for a preliminary look to see what they had. First they needed to correct for the machine's emitter-detector misalignment, which had resulted in images that were stretched one way and compressed another. The data was good, the images were just skewed, and the process to correct the error was relatively easy, if time-consuming, without certain xyz coordinates Pruitt didn't have access to. Pruitt and Schlader sat side by side for three days, with Pruitt measuring the actual specimen and relaying the numbers to Schlader, who would plug them into the computer to shift the image slightly, first one axis, then the next. Once the scans were properly aligned they ran a "first calculation," and up popped what looked like a roadkill duck head. A beautiful, beautiful roadkill duck head. It was jaw material, no doubt about it. Ironically enough, the tooth whorl itself didn't show up on the scans because it was more impression than preserved material. Pruitt and Schlader generated a crude 2-D computer rendering of the duck head, which convinced them the scans were a big success and had captured definitive evidence that—if solidly interpreted and properly presented—could resolve the hundred-year-old debate of exactly how the whorl fit in *Helicoprion*'s mouth.

Feeling confident, they were anxious to see the more highly refined scans and models that the team in Austin would produce—except cost estimates for the processing work came in beyond their depleted budget and out of reach. They would have to process and model the scans themselves. Rather, Pruitt would have to do it, with Schlader's help. Schlader was experienced with CT-processing software, although none of that experience involved interpreting rock-imbedded fossil scans. CT scans work off density variations, so high-density materials appear white, and low-density materials appear black. That works nicely in a CT scan of a human arm, or of an Ice Age camel femur collected from the mud at American Falls Reservoir—the bone shows white, and everything else is a highly contrasting dark gray to black. The problem with a fossil, however, is that the fossil and the rock it's imbedded in, the matrix, can be nearly the same density, so will show very little variation. Fossils in a sandstone matrix might have a decent range of densities, but in Idaho no. 4, the contrast between the fossilized cartilage and surrounding phosphatic rock was weak.

Previous to his involvement with *Helicoprion*, Pruitt hadn't had much interaction with IVL. When the lab moved to the museum and earth sciences building from another location on campus, Pruitt had helped move books out of the room they came to occupy, an old department library. Then after the move he occasionally dropped by to visit Schlader and assistant manager Nicholas Clement to see what interesting things they were up to, but that was the extent. Now he was spending eight hours a day in that room, with its computer screens, digitizing equipment, metal storage cabinets, and imaginative IVL-created paleo-Comic-Con-style posters featuring everything from elegant compositions of whale skeletons and ancient Arctic harpoon tips to whimsical, fantasy realm flute-playing half-human, half-horse skeletons.

Pruitt had no prior experience with graphic or digital processing software, but he did have a lot of experience by then in traditional fossil prep, working on physical specimens to methodically remove rock matrix to release the buried fossil. Fossil prep is a series of careful steps, which when done with great care and attention to detail, result in a sound, scientifically useful fossil. Pruitt found "digital prep" to be similar, with its careful, methodical steps inherently and necessarily guided by the material at hand. He turned out to be a natural at the computer work.

Schlader taught Pruitt how to exploit their medical CT processing software to extract as much information as possible from the scans. To do that, Pruitt would bring up one of the 360 individual slices to his computer screen and manually, digitally, "paint" any cartilage he detected in the scans. So, in a grainy gray block, wherever he saw a line or area of gray that was slightly different than any other gray, he would paint that in, a process known as masking. Schlader also showed him how to use thresholding, an auto-masking feature that allowed the operator to set parameters for varying densities. With thresholding, the computer can automatically reveal areas of subtly differing densities. But the gray scale on their scans was so flat because of the density of the rock, Pruitt was relegated to a mostly manual masking process. Just as he had measured, measured, measured tooth crowns in the basement with his calipers, he sat in front of a computer screen to mask, mask, mask—three views per slice, starting with the front-to-back view, then moving to the top-to-bottom view, then the lateral, or left-to-right views, which had the most compressed information. Once that was done he would compare the three masks, find his bleeds, and clean those up. He did that one by one for every slice, which on his half-time schedule took about two months. Maybe it wasn't all that different from hunching over a rock scraping away with a dental tool for weeks on end. It's all in the name of prep, and it trained his eye for reading scans.

Once Pruitt had the scans processed, he and Schlader were ready to generate a new, cleaner 2-D rendering and build a solid 3-D computer model from there. Three-dimensional models are important because they allow scientists to animate and manipulate a reconstructed fossil to ferret out morphological and mechanical details, such as how a certain joint might work. Three-dimensional computer models can also be printed out as actual physical models through the rapidly evolving 3-D printing process. UTCT was an early adopter of 3-D printing as well as CT technology. In the early 2000s, British Museum curator Angela Milner carried one of the world's most precious fossils to UTCT wrapped in tissue in a box hidden in her shirt pocket—the brain case and inner ear of *Archaeopteryx*, an evolutionary intermediary between birds and dinosaurs. The CT scan took thirteen hundred slices of the tiny fossil, assembled them into a 3-D computer model, then printed that out in wax, with every detail intact. The images clearly showed blood vessels and inner ear canals, revealing clues about the creature's sense of balance and hearing.

Pruitt and Schlader had their computer model, but it was cluttered with broken cartilages and hard to interpret. What was what, and which went where? It was like they had the Dead Sea Scrolls but didn't know how to unwrap and read them. They sent the computer file to Didier, who had recently been to the American Museum of Natural History to see her colleagues John Maisey and his French paleontologist postdoc, Alan Pradel. Pradel had studied under the esteemed Philippe Janvier, and was now putting his exceptional skills in reading and interpreting CT data to work for Maisey, a leader in using scans to understand the anatomy and evolutionary relationships of the earliest chondrichthyans. Pradel had already established his own reputation by noticing a ghosted object in scans he did of the braincase of a three-hundred-million-year-old iniopterygian—the carnivalesque early holocephalan first identified by Rainer Zangerl from the Mecca shale. Pradel's iniops had been collected in Kansas in the 1960s, from a quarry now paved over by a Walmart. He scanned it again and realized he was looking at a fossil brain the size of a pea. At the time, it was the oldest fossil brain ever detected. (In 2015, scientists published on an even older fossilized brain, of a trilobite ancestor more than five hundred million years old.) The fossilized iniopterygian brain was similar to the brain of a modern ratfish, in that it was notably smaller than the braincase itself. In many so-called "lower" vertebrates—vertebrates such as fish and amphibians that lay eggs in water, as opposed to "higher" vertebrates like chickens and humans that produce amniotic eggs—brains quit growing while skulls continue to enlarge. While Didier was in New York, Pradel had shown her some of his amazing models.

Troll had also corresponded with Pradel, and he and Didier agreed that input from someone in the Paleozoic shark world like Pradel would be enormously helpful. In August 2011, Didier emailed Pradel to gauge his availability and interest in working with the group. Pradel was intrigued by the possibility that there might be new information on *Helicoprion*—maybe even cranial material, at least according to Bendix-Almgreen—which in turn could shed light on the broader group of Paleozoic chondrichthyans he and Maisey were investigating. Maisey approved some extracurricular time for Pradel to help out, so Pruitt sent along their raw data for him to run through his more powerful software and computer. Pradel's first reaction to the data: "Tu plaisantes, n'est-ce pas?

You're kidding, right? This is horrible." Then Pruitt sent the files he had masked. (*"Okay, better."*) Once Pradel saw what they had been calling out, he went back through the slices and found even more pieces of preserved cartilage than Pruitt had. His input was as valuable as they hoped it would be. In less than a week, Pradel had sent back a new 3-D model with better resolution of the fossil's features. It was still a mess inside the concretion, he noted, with lots of broken cartilages, but he identified the elements he could, pointing out parts of the upper jaw (palatoquadrate) and lower jaw (Meckel's cartilage). Despite looking as hard as he could, however, Pradel could not find evidence of his favorite fossil bit, the cranium. Along with the model, Pradel sent Pruitt useful tips on pulling images into Photoshop to adjust contrast, and offered advice on thresholding and other ways to maximize the utility of their software. With Pradel's data, interpretations, and tips in hand, Pruitt went back through the scans one more time, slice by slice, to squeeze out every last ounce of "sharky goodness."

The resultant new model was many steps beyond roadkill, which opened the floor to a fresh set of questions. Didier wondered how the teeth returned to the lower jaw. Tapanila wondered what accommodation the jaw might have for the older parts of the whorl. Pruitt wondered about the mechanics of the jaw. Pradel wondered if the mouth opened at the front of the head, as it did with a number of ancient chondrichthyans, rather than opening more on the underside of the head like a modern mako shark. Troll wondered whether the shark used its tooth whorl to slash prey or grab, crimp, and rip!

With monster sharks tearing freshly through his imagination, Troll had already floated the idea of creating a *Helicoprion* exhibit for IMNH, since they were, after all, the world's leading repository of *Helicoprion* specimens, with more whorls than any other single institution on the planet. Then–museum director Herbert Maschner had given preliminary approval for the exhibit, and now, with the growing momentum of Pruitt and Tapanila's work and the excitement of Pradel's model, it seemed like an even better idea.

Although Troll was already calling them "brothers of the whorl," he, Pruitt, and Tapanila had never met in person. That changed in November 2011, in a confluence of people, place, and circumstance around the annual meeting of the Society of Vertebrate Paleontologists (SVP), in Las Vegas. Troll was going to be in the area for a gathering of his wife's family, Tapanila would be there on his way back to Idaho from a field trip, and Pruitt was driving to Las Vegas to pick up the fossil he left in Texas, which the UTCT folks were bringing up for him.

They found one another on the sidewalk by one leg of the faux Eiffel Tower—Pruitt, Tapanila, Troll, and paleosculptor Gary Staab, who had been cruising the SVP venue with Troll. Staab of course had his own *Helicoprion* connection, having visited the museum years before at Troll's suggestion and cast one of the whorls in bronze while he was there. Spirits were high as the group retreated into the Paris Las Vegas Hotel and tucked around a tiny wicker table at a "sidewalk café" in the hotel's sprawling interior, built like a theater set of the streets of Paris. Adult beverages poured in as conversation and a special camaraderie poured out. Nearly fourteen hundred people attended the SVP meeting that year, with celebrity paleontologist and popular speaker Jack Horner giving a keynote on "Dinosaurs and the Proofs of Evolution." In the whole field of vertebrate paleontology, if you shuffle all the dinosaur workers over here, put the marine reptile people over there, and shunt the mammal folks off yonder, what you have left is a relatively small pool of fish people. If you siphon off from that the researchers actively working on Paleozoic sharks, you're lucky to get enough people for a basketball team. No one at SVP knew it yet, but the team was set to expand. The new friends talked Paleozoic sharks as the evening progressed and the din around them grew. It was the night of the big SVP dance party, and the driving electronic beat drew them to the dance floor like sharks to fat seals. Working the crowd as always, Troll goaded Pruitt and whoever else would bite into creating and demonstrating a special *Helicoprion* dance move, the "chop-saw-whorl-chop."

Early the next morning, Troll stumbled out to meet Pruitt in the back parking lot of Caesars Palace to see the big whorl he was taking back to Idaho, which had proven too bulky to scan. Pruitt perched the hundred-pound chunk of ancient shark on a fleece camouflage jacket on the trunk of his car, spitting distance from the hotel's massive, graffiti-scribbled Dumpster. Even though only a wedge of the whorl remained, it was a big wedge, and IMNH 30897 was a stunning specimen, from its exquisitely delicate juvenile tooth arch, through four and a quarter spiraling turns, to impressions of teeth the size and shape of Stone Age spear points. While Troll ogled the fossil, people wisped by, part of the Las Vegas twenty-four-hour flow. A car pulled in next to them and two girls in short skirts and high heels clutching go-cups emerged trailing boyfriends. They were drawn to the fossil but held slightly back, instinctively reverent, or maybe just nervous to see something so old. Pruitt and Troll told them about *Helicoprion*, then Pruitt gathered up the fossil, resettled it into the backseat, and headed north.

Back in the saddle in Idaho, Tapanila and Pruitt pressed on with their research into species characteristics. Troll was anxious to move forward on the exhibit, or anything else that would push *Helicoprion* into the limelight, but Pruitt and Tapanila were buckled down to the research questions in front of them. Now working together outside the box of teacher and student, they had shifted gears into a productive, egalitarian research partnership driven by their collaborative synergy and the conviction that important answers were inching into reach. Tapanila approached the work mathematically, Pruitt's perspective was mechanical and spatial. In wide-ranging discussions, Tapanila could throw out ideas, and Pruitt could tell him why he was right, or why and how he was wrong.

Feeling restless, Troll flew to Idaho in January of 2012 to talk up the exhibit with Maschner and push Tapanila and Pruitt on the scanning and modeling work. During his visit, Troll hung out with Pruitt and Schlader in the IVL, watching them experiment with their animations of the *Helicoprion* jaw model, making it open and close. How far would it open? Where would it hinge?

You guys need someone who knows about modern shark jaws, Troll said over their shoulders. Having gotten the official go-ahead from Maschner on the exhibit, Troll was in full-steam producer-and-director mode. He had the big picture; he knew the players; he wanted to help. *A jaw expert. That's what you guys need.*

While CT scanning had become the sexy, ultramodern fashion in paleontology, the practice of comparative anatomy was still the essential sensible shoes. You had to walk that comparative path to get anywhere in understanding extinct animals. Pradel, for instance, in his work on ancient iniopterygians, had turned to Didier for her expertise and dissections of modern chimaeras. Tapanila and Pruitt needed an expert on the jaws of living sharks. Of course, Troll knew just the person. She was Alaskan.

12

COMING TO TERMS

Sharks have everything a scientist dreams of. They're beautiful—God, how beautiful they are! They're like an impossibly perfect piece of machinery.
—Peter Benchley, *Jaws*, 1974

CHERYL WILGA MIGHT HAVE JUST COME BACK FROM CHECKING ON HER BAMBOO sharks when her office phone rang. It was January 2012. Wilga taught biological sciences at the University of Rhode Island, and maintained a small colony of bamboo sharks for her research into the feeding behavior and jaw mechanics of modern sharks. Bamboo sharks are well-suited for aquarium life—the attractive, easygoing creatures are about five feet long at their full size, and can be trained to eat out of your hand. They have an endearing habit of resting on the bottom of the tank propped on "elbows" they make of their pectoral fins, like they're lying on the beach watching the scenery. Contrary to popular shark mythology, not all sharks have to swim constantly to survive. Many species, including bamboo, lemon, bullhead, and nurse sharks, commonly rest on the seafloor, keeping oxygenated water moving over their gills by slightly opening and closing their mouths. (The gill slits of "benthic," or bottom-living sharks are typically located just above their pectoral fins, possibly to avoid stirring up too much silt when the gills are working.) Bamboo sharks are suction feeders, meaning that instead of biting or grabbing prey, they catch a passing shrimp, small fish, worm, or crab by sucking it into their mouth. The hapless victim doesn't know what hit it, as the whole thing happens in a split second of impressive vacuum force, like the intake stroke of a bellows. By implanting metal markers into the jaw and surrounding cartilages of a bamboo shark, Wilga and her colleagues had produced astonishing X-ray videos showing the skeletal mechanics of suction feeding in live sharks. You have to watch it in slow motion or you miss the whole show.

Given our American obsession with sharks—most precisely the yawing, bloody, tooth-studded jaws of sharks—you would think the mechanics of how those jaws work would have been studied to death. You would be wrong. Barely on the other side of fifty years old, Wilga was one of the world's few authorities on the subject. The small, cheerful, dark-haired woman was among an even more elite contingent that studied the way jaw suspension (remember the cranial suspenders?) affected the mobility of shark jaws, which in turn affects feeding style. She earned her PhD from University of South Florida, with a dissertation on the "Evolution of Feeding Mechanisms in Elasmobranchs: A Functional Morphological Approach." Wilga completed post doc work at the University of California, Irvine, and Harvard, and had been at the University of Rhode Island since 2000.

Troll didn't know Wilga personally, but he knew her work, first learning of her and her lab in a 2006 article in *Natural History* magazine. The article, "When the Shark Bites," was written by Adam Summers, an engineer/mathematician turned biologist/artist/writer and scientific consultant on the animated Pixar films *Finding Nemo* and *Finding Dory*. Summers's article was about a "shrewd observation" made by Wilga's graduate student, Jason Ramsay, that the teeth of bamboo sharks fold down when the shark bites hard-shelled prey, turning sharp cutting teeth into crushing plates.

Wilga likewise knew of Troll. While living at the Coast Guard Air Station Kodiak in Alaska as a young bride, Wilga had run the base exchange liquor store, which she transformed into a wine shop. Wilga knew retail, and in Alaska, retail and Ray Troll T-shirts went together like hot dogs and mustard. When she picked up the phone that day in her office, she thought someone was playing a joke on her. The Alaskan artist and fishy cult hero Ray Troll on the phone? Why? Wilga listened with interest as Troll described the *Helicoprion* work going on in Idaho. They had the fossils, he said, they had the scans, they had the computer models, now they needed to put some bite into it all.

Although Wilga worked exclusively with modern animals, she had scouted the world of paleo sharks in her investigations into jaw evolution. In 2008, she had been invited to participate in an international symposium of paleontologists as one a few biologists working on extant animals, so she knew John Maisey, Eileen Grogan, and some of the other prominent paleo people. She had never heard of Tapanila, who came up in the invertebrate world, but she did know about *Helicoprion*. Wilga's innate curiosity had driven her into research; she was one of those kids who took things apart to see how they worked. Now here was Ray Troll asking if she would help figure out how the fantastically odd business end of this most baffling beast operated. She took the bait.

Tapanila and Pruitt were somewhat dubious about what Wilga could bring to the table, but they would find out quick enough. Besides her academic credentials, research experience, and rare expertise, she brought the energy and determination that she applied to everything in her life, just like her mother before her. Wilga's mother is a Dena'ina Athabascan Native from Alaska's Kenai Peninsula, who as a child, had lost her mother and two siblings to tuberculosis and was left to raise three surviving sisters. Still, she managed to convince her father to let her go to school, and in a fortunate turn, ended up at Mount Edgecumbe in Sitka, one of the better-run boarding schools operated by the Bureau of Indian Affairs. The resilient girl thrived at the school, graduating as valedictorian of her class. She won a scholarship to the University of Alaska–Fairbanks, where she met and married a handsome French-Blackfoot air force serviceman from upstate New York. Cheryl was born while the family was stationed in Presque Isle, Maine. Over the years, the family lived up east and in California, Hawaii, Wyoming, and Massachusetts. After high school, Wilga attended Springfield Tech to study computer systems. Right about the time she was thinking a life in computer systems wasn't for her, she married her high school boyfriend, who had enlisted in the coast guard, and moved with him to Kodiak Island. This daughter of the Dena'ina had never even been to Alaska before, but Wilga felt like she was home at last. The air, the mountains, the ocean—somehow she had always been there. So much was so right in Kodiak, but like her mother, Wilga felt an irresistible pull to advance her education. After receiving her associate degree in biology from the University of Alaska–Kodiak (as valedictorian, again like her mother), she knew she couldn't stop there. Wilga moved with her husband, two cats, and a dog to Florida to attend the University of South Florida, where she met Philip Motta, a pioneer in the study of shark feeding behavior. She began helping Motta determine which muscles were activated by what events in different species, and thus discovered the rarified world of jaw suspension.

In humans, the upper jaw is fused to the rest of the cranium. In other words, the upper jaw can't flop down from the brain case. As you can prove to yourself right now, the lower jaw is the moveable part, connected to the upper jaw/cranium by joints, muscles, and ligaments. As a human, that's all you get. Sharks, on the other hand, have a number of styles and configurations for attaching jaws to the cranium—more than any other vertebrate. That's fitting because jaws first evolved in fish, and sharks are among the oldest fish, so have had plenty of evolutionary time to expand their suspender collection.

Jaws appeared before teeth, in the Silurian period, which began around 444 million years ago. At that time, all the animals on earth, save for an innovative group of centipedes, millipedes, and arachnids, lived in the water. Scientific tradition holds that jaws initially evolved to help early jawless fish breathe. Fish need oxygen to live, just like we do, but instead of inhaling air into lungs, they take up dissolved oxygen through their gills. A mechanical contraption allowing the mouth to open and close with minimal energy could create a steady current of water over the gills to help fish breathe easier, especially in poorly oxygenated waters. Evolution is supremely complex and multifaceted, but the kindergarten CliffsNotes might say that at its core, evolution is the repurposing of existing parts. Jaws are a great example. Early jawless fish had a series of gill openings behind their mouth, supported by cartilaginous arches. It's thought that the gill arch nearest the mouth nudged its way forward and adapted into a hinged jaw. Improved oxygen intake led to increased mobility, which opened up new possibilities for feeding, which drove further evolutionary changes. A few jawless fish, like lampreys and hagfish, hang on today, but in an evolutionary backwater as parasites and scavengers.

The next big evolutionary coup after jaws was the teeth with which to outfit them. Although the origin of teeth is "mired in controversy," there are two main theories: outside-in, and inside-out. The outside-in theory holds that the first tooth to set foot in a jaw migrated there from its place as an external dermal denticle. Denticles, we already know, have a toothlike structure with pulp and dentine. Early jawed fish had hardened denticles bearing small nubs or barbs on the rims of their mouths, and it's thought those denticles worked their way into the mouth and became teeth.

The inside-out theory is a little harder to swallow. It holds that teeth started in the pharynx, the muscular tube at the back of the mouth that opens onto the esophagus. Certain ancient fish had jaw-like structures situated in the pharynx, called "pharyngeal jaws." It might surprise you to learn that some thirty thousand species of bony fish living today also have pharyngeal jaws, which are often lined with pharyngeal teeth. Goldfish have them, as do moray eels. Perhaps that was what a couple of Smithsonian paleontologists were thinking in their 2008 reconstruction of *Helicoprion,* which placed the whorl down in the shark's throat. That reconstruction, although loyally and vigorously defended until 2015, was a head-scratcher, and only partially because pharyngeal teeth are generally on the blunt side, most suited for grinding or shredding. It's true that some pharyngeal teeth are sharp and designed for grasping, but those are found in more evolved bony fish, so are unlikely to have any relation even to a similar looking structure in chondrichthyans. But anyway, in the inside-out scenario, genes got to work instructing those pharyngeal teeth to move to the jaws and become oral teeth. Recent papers have argued that the inside-out hypothesis "must be rejected," but the debate continues.

With jaws and teeth squared away, evolution set out to test different options for attaching the jaws to the cranium—with sharks, as mentioned, the undisputed champions of jaw suspension diversity. The issue of jaw suspension is very significant to the *Helicoprion* story, as you'll soon see.

For modern sharks, the most common type of jaw suspension is called "hyostyly." In hyostyly, the upper and lower jaws, which together are called the "mandibular arch," are reinforced at the back with what's called the "hyoid arch." The hyoid arch is the second-in-line repurposed gill arch, which evolutionary forces moved forward to tack onto the mandibular arch like seam binding tape. The top section of the hyoid is free, like the end of the seam tape has come loose. This free part of the hyoid is the main attachment point to the cranium. Monkeys are a big leap from seam-binding tape, but the hyoid attaches to the cranium like a monkey holding on to a tree branch with one arm, leaving its legs free to maybe grab a piece of fruit. In hyostyly, ligaments also connect the upper jaw to the cranium in the ethmoid region toward the front of the brain case, but the hyoid is the primary skeletal and most positive point of attachment. This minimalistic suspension allows the entire mandibular arch to sling forward and open wide when the shark goes to take a bite of seal or surfboard or whatever it's chomping. Along with hyostyly, modern sharks can also have amphistylic, orbitostylic, and euhyostylic jaw suspension, all of which incorporate a hyoid arch attachment plus a certain design of ligament attachment.

Modern holocephalans, on the other hand, have one, single jaw suspension system, called "holostyly." In holostyly (*holo* for "whole, entire"), the upper jaw is fused to the braincase. As in humans, the upper jaw is one with the cranium. This style robs holocephalans of the ability to protrude their jaws, but in trade it gives them a beefy setup to support their crushing tooth plates and durophagous habit of consuming hard-shelled prey like clams and crabs. At the other end of the spectrum, skates and rays have "euhyostylic" suspension (*eu* means "true," true hyostyly) in which the jaw is suspended only by the hyoid arch, with no ligament attachments at all. This gives them the most license of all chondrichthyans to protrude their jaws, which makes a lot of sense for flat fish, and can even help bottom feeding species create suction to pull prey up from the substrate. You begin to see how jaw suspension and feeding behaviors correspond.

Jaw suspension wasn't on Troll's radar when he called Wilga. He just knew that having a biologist knowledgeable about the physiology of modern sharks would move the *Helicoprion* work forward. He remembered something Didier had told him years before, about Eileen Grogan and Dick Lund. Lund had been excavating Carboniferous chondrichthyans from the Bear Gulch quarries in Montana for years when he first met Grogan. She was visiting Adelphi University to see about pursuing graduate studies there. At the time, Grogan was a "white lab coat" biologist working in Manhattan, interested in cancer research. The scientists she had come to Adelphi to see told her the theory that sharks don't get cancer, and sent her down the hall to talk to Lund, who was the campus shark expert—albeit fossil sharks, and not exactly "sharks," but ancestral holocephalans. Grogan ended up getting her masters in biology at Adelphi and her PhD in marine science from the College of William and Mary, writing her dissertation on "The Structure of the

Holocephalan Head and the Relationships of the Chondrichthyes." As part of her work, she studied the blood and cells of living sharks, falling in love with sharks, then chimaerids, then Lund.

She became curious about how fossil and modern chondrichthyans were related, and subset to that, how elasmobranchs and holocephalans were related. How were they the same, and how were they different? As she pulled the thread on her questions, she gradually transitioned from studying living forms to investigating fossil forms. One day, while looking at a fossil fish Lund had published on years earlier, Grogan said, "Oh look, that fish died of asphyxiation." The fossilized gills were flared out, a sign that before its death the fish's gills had been engorged with blood, a symptom of suffocation. That detail provided insights into the environment in which the fish lived and the circumstances under which it died. Fossil CSI. The story had made an impression on Troll, and was why he thought a biologist could parse the scans and fossils for lifestyle information on *Helicoprion*. Specifically, a biologist specializing in jaws. Most specifically, Cheryl Wilga. After the exploratory call to his fellow Alaskan, Troll was excited to tell Tapanila and Pruitt they had their next team member.

Reserving their own judgment on that, Pruitt and Tapanila sent Wilga the computer models they had refined with Pradel's help, and Troll set up a Skype call to take place before he headed back to Alaska. Wilga invited her thirty-nine-year-old graduate student Jason Ramsay, the "shrewd observer" of folding bamboo shark teeth, to sit in on the call. Ramsay had been pursuing a career in art until he took an undergraduate biology class with Wilga and discovered the field of functional morphology. At that time, Ramsay had been a father of four, working a full-time job until midnight along with taking classes—and still managed to be the top student in Wilga's comparative anatomy class. Now in the final mile on his PhD, Ramsay had become a single father of four, and was working as a part-time teaching assistant at the University of Rhode Island and full-time night-shift custodian at a middle school, in tandem with his academic work. Functional morphology examines the relationship of an organism's anatomical form to its function, like how the shape of a fin or a foot or a pelvis or a tooth influences how an animal moves or eats or lives. The field was a perfect fit for Ramsay, who, with his linebacker build, backward baseball cap, and disarming smile, looked like he could make himself fit anywhere he felt like. Ramsay remembered seeing *Helicoprion* on Shark Week, and couldn't believe he was getting such a close-up look at the crazy creature.

If Tapanila and Pruitt had any lingering doubts, they evaporated on that first Skype call as Wilga and Ramsay identified potential muscle attachment points and other features. Everyone was busy with teaching, work, studies, and life, so the long-distance *Helicoprion* conversation stretched out over a number of months. Wilga identified the upper from the lower jaw, and she and Ramsay studied the scans and computer models, asking questions as they came up, and sharing observations, information, and ideas with Tapanila, Pruitt, Didier, Pradel, and Troll. The Idaho group, consumed as they were by the CT scan discoveries of jaw material, had all but forgotten about the tooth whorl. Not so Wilga and Ramsay, who hadn't been up to their eyeballs in whorls like Pruitt and Tapanila had been. The Rhode Island folks wanted teeth to go with their jaws. Since the tooth whorl in Idaho no. 4 was almost entirely an impression, not fossilized material, the whorl had not been captured by the CT scan. So Schlader set about to digitally construct a *Helicoprion* whorl to plug into the computer jaw model. While he built the whorl, he had epiphanies that he shared with Pruitt, who was sitting at the next monitor over, since Pruitt was now working as an intern at the IVL. Foremost was Schlader's earned understanding that *Helicoprion*'s tooth whorl was one, single tooth—one root with multiple crowns. Many sharks have multiple tooth families, that is, teeth in different positions on the jaws have different shapes. Not only did *Helicoprion* have only one tooth family, it technically had only one tooth. When Schlader started his process, he had been trying to stack together multiple individual teeth, and it wasn't working. Once the anatomical truth sank in, he spun out a beautiful logarithmic whorl.

The process led Pruitt and Schlader to think more deeply about the tooth growth pattern in general. *Helicoprion* is unique in the shark universe for its full-circle tooth spiral, in which new crowns erupt, while older ones are reeled in instead of being shed. In a sense, though, all shark teeth have a spiraling growth pattern. Shark teeth aren't rooted into the jaw like mammalian teeth, but are formed inside a membrane in the jaw. As each tooth grows and develops, like in great white shark, the membrane rotates it ahead as the replacement tooth comes up behind, not in a back-to-front conveyor belt progression, but in an inside to outside roll—imagine lemmings running over a berm, where the lemmings are the teeth and the berm is the shark's jaw. Generally, each replacement tooth is larger than the one that preceded it, matching the animal's overall growth in body size. Teeth emerge at the size they will be forever. They don't continue to grow after they've erupted. *Helicoprion* may not have shed its crowns as it generated new ones, but they were still growing along a curve.

Schlader scaled the digital whorl to fit Idaho no. 4, and finally had a file to send to Wilga and Ramsay, who pored over the puzzle before them.

For those who ponder *Helicoprion*'s mysteries most intensely, there seems to be something of a rite of passage—a dizzying, disorienting moment when the fossil swims before their eyes, and what made sense a second before suddenly doesn't anymore. That came for Ramsay when the shark turned on him. Pradel and Didier had already identified in the scans which was the upper jaw, or palatoquadrate, and which was the lower jaw, called the Meckel's cartilage. It may seem unnecessarily complicated to call an upper jaw a palatoquadrate, and a lower jaw a Meckel's cartilage, but one of the terrible beauties of science is the precision of its language. In the sphere of chondrichthyan anatomy, a jaw is not just a jaw—it's a mandibular arch, comprised of a palatoquadrate and a Meckel's cartilage. In humans and dogs, it's maxilla and mandible. The precisely selected vocabulary word holds within it context and extra detail. An exact word forces intellectual attention and can contribute to clarity of thought. That said, there's always room for personal expression. Tapanila, being the good Canadian he was, thought the Meckel's was shaped a bit like his hockey stick, and sometimes called it that for variety. The whorl went with the hockey stick.

At one point, Ramsay noticed a double joint near the back of the other, hatchet-shaped, jaw piece. The rear portions of both jaws on Idaho no. 4—the areas where upper and lower would have been jointed together—had been well preserved, and the team zoomed in for a better look. In modern sharks, that kind of double joint was only present in the lower jaw. It's what kept the jaw from dislocating when a shark shook its head from side to side to rip apart a seal, for instance. Ramsay fell dizzily down the *Helicoprion* rabbit hole. Was the hockey stick really the upper jaw? But that would mean the whorl was in the upper jaw. *You guys? Do we have this upside down?*

Stomachs turned in Idaho, and in Alaska, Troll gnashed his teeth remembering 1995, when he was installing his first "Dancing to the Fossil Record" exhibit at the California Academy in San Francisco. A scientist acquaintance big in the shark world dropped by, looked at Troll's *Helicoprion* drawing, and said, *Nah, shark teeth don't grow like that. You have the whorl in there backward. You should have the biggest teeth at the front.* The exhibit was about to open at a prestigious science institution, and a famous shark scientist just told Troll he had it backward. J. D. Stewart had brought the L.A. fossil up for the show, and Stewart talked Troll back from the ledge over lunch at a sushi joint near Golden Gate Park. The newer teeth are the bigger teeth, and belong in the back. *You have it right, man.*

Now here was Ramsay, potentially turning the work on its head once more.

The agitated group went back and forth on the issue, as they scrambled to sort it out. They ultimately agreed that the hatchet-shaped piece was the palatoquadrate, as Wilga had first said, because it had clear anatomical features like articulation points and processes that were associated with upper shark jaws. And they agreed the hockey-stick-shaped piece was the Meckel's, the lower jaw, because that piece clearly held the whorl, and most tooth whorls in closely related species were in the lower jaw. The team had its moment of doubt, a great tradition among *Helicoprion* workers, forged through it, and moved on.

As unnerving as it was, hammering through the question to a consensus answer cohered and encouraged the group. Energy was high, and Tapanila started to feel like he had a tiger by the tail. They began to talk about collaborating on a paper, and decided to meet in Pocatello to decipher the rest of the interpretation together. Troll was thrilled. He had been fomenting for just such a meeting—a Shark Summit! His shark was finally going to get the attention it deserved.

By then it was September 2012, and the summit was quickly planned for early October, to include Tapanila, Pruitt, Schlader, Wilga, Ramsay, Pradel, and Troll. Team Helico. (Didier had to teach, so she couldn't go.) Pruitt and Schlader fine-tuned the computer model with all they'd been discussing over the previous months, while questions, observations, updated models, and ideas continued to fly via email.

Just FYI team, wrote Pruitt, *the lower jaw is around thirty cm [11.8 inches] and the whole critter scaled to the preserved material comes out to just under fourteen feet! He was a pretty big fish, and this fossil is less than half the size of our biggest specimens!*

Troll began sketching a shark around the scan models, experimenting with nose length and buzzing the group with questions. *Is there enough room for the whorl to work properly without a long nose? Would a longer nose give a shark more muscle attachment points, or larger "lifting and biting" muscles? Is there a hydro-dynamic advantage for a long nose?*

As always, Troll's illustrations would ultimately be an interpretative best guess. He wanted it to be the best of best guesses, but at the same time, he was hoping for his monster shark. When Pruitt and Schlader set the number of teeth showing in the mouth at around ten, Troll pushed back for more. Pruitt said the estimated fifteen to seventeen teeth he and Tapanila had previously suggested were based off a specimen that, as it turned out, may have been a "freak among freaks." Pruitt explained his and Schlader's logic: *eleven exposed teeth is the best compromise of no wasted material around the Meckel's cartilage, an effective cutting blade, and aesthetics.*

Art is in the details, and Troll was prepping for his part of the summit—bringing *Helicoprion* to visual life. He wanted to know if

Wilga thought *Helicoprion* could have had a nictitating membrane, the membranous "third eyelid" that some animals have. He might be able to use that to effect in drawing the eye. Polar bears and bald eagles have nictitating membranes, as do the Carcharhiniformes, Wilga noted, the largest order of living sharks, which includes hammerheads, tiger sharks, blacktip reef sharks, and dogfish. Lamniformes—including great whites, thresher, and goblin sharks—didn't have them. Those sharks probably rolled their eyes for protection when snatching prey.

Anticipating his own primary line of inquiry for the summit, Pradel brought up the question of jaw suspension. In an email to the group, he asked if they had inferred a certain type of jaw suspension from their reading of the scans. Had they been mulling over the phylogenetic significance of discerning that detail? (Phylogeny being the evolutionary history, the development and taxonomic relationships, of lifeforms.) He certainly had. The rest of the group, however, had been most focused on whorl placement and function. Being recent initiates into the paleo shark world, they weren't as fully cognizant of just how murky the realm of chondrichthyan phylogeny was, and therefore how significant it would be to determine the jaw suspension type of a Paleozoic species, especially one of the eugeneodontids. Pradel's emails crossed at the same time Wilga was asking Schlader about the possibility of showing throat expansion in the computer model, based on the action of the hyoid arch. Well, said, Schlader, there doesn't seem to be a hyoid arch. At least it wasn't in the scans. There had to be a hyoid, said Wilga. All sharks have a hyoid to support the jaw and expand the throat area.

For some people, Pradel jumped in to explain, Helicoprion *is not a shark (elasmobranch), but it may have been rather some kind of primitive chimaeroid (holocephalan).* Pradel went on to explain that early holocephalans, like modern chimaeroids, do possess a hyoid arch, but that arch doesn't support the jaw, as it does in elasmobranchs. Instead, the arch supports the operculum, the gill cover. Troll responded that he knew Dick Lund and Eileen Grogan were also of the opinion that *Helicoprion* was an ancestral holocephalan— but he resisted the notion, because in his artist's mind, he was lumping all early and ancestral holocephalans into ratfish-looking creatures. With his angst leaking through, he wrote, *As much as I dearly love chimaeras I would be very surprised to find this critter over on the holocephalan side of the tree. It would radically alter the look of the animal*—the animal in which he had twenty-plus years of artistic investment. It was another one of those dizzying *Helicoprion* moments. Protesting that *Helicoprion* couldn't have been a holocephalan, he listed *Ornithoprion, Sarcoprion, Caseodus,* and other sharky-looking eugeneodontids pictured in Zangerl's 1981 *Handbook* that were closely related to *Helicoprion.*

What Troll didn't realize was that the tiny set of paleoicthyologists working in the weeds on the phylogenetics of paleo sharks had moved away from Zangerl's lumping of eugeneodontids into the elasmobranch group. Those few who cared and who made a study of the question, like Lund and Grogan, had made the convincing case that the Eugeneodontida order belonged in the Holocephali subclass of class Chondrichthyes, as a sister group to the Elasmobranchii subclass. Adding to the confusion, the term "holocephalan" had come to be interchangeable with living chimaerids, the real ratfish, and their most recent ratty-looking ancestors from the Mesozoic. That's one of the reasons Lund and Grogan created the term "euchondrocephalan" to refer to ancestral holocephalans in a way that would gather in the wildly diverse and abundant Paleozoic chondrichthyans that didn't fit into the elasmobranch group, but that still didn't look like ratfish—the so-called stem holocephalans, like *Sarcoprion,* et al. Just that year, John Maisey published a paper titled "What is an 'elasmobranch'? The impact of paleontology in understanding elasmobranch phylogeny and evolution." Classification of early chondrichthyans is all over the map, and as such, Maisey proposed starting with something of a clean slate and considering "elasmobranchs" as the equivalent of neoselachian or "modern" sharks, by which he meant sharks that arose following the Paleozoic, including megalodon. That would neatly slice away the problematic ancestral holocephalans and Paleozoic sharklike chondrichthyans like the eugeneodontids. The elasmobranch group had been erected in 1838, and for a very long time served as a big bucket into which sharks, rays, and chimaerids—so essentially all the chondrichthyans—were collectively tossed. Not only had a world of fossils been discovered since then, but also galactic advances had been made in the sciences of paleontology, genetics, phylogeny, and morphology. Maybe it was time to dump the bucket and re-sort the contents.

In any case, Pradel wasn't saying that *Helicoprion* wasn't sharklike. Pradel's idea that *Helicoprion* might possibly have been an ancestral holocephalan was rooted in a detail of jaw suspension, not necessarily body shape or size. Pradel was looking at the inside of *Helicoprion,* and Troll was looking at the outside.

Pradel was generously understated in his emailed response: *Indeed, the phylogenetic position of Eugeneodontids is still problematic.* Whatever they would end up calling it, Pradel said he agreed with Zangerl's original advice to Troll that it would be reasonable to model *Helicoprion*'s body shape on *Caseodus.* Just because *Helicoprion* may have had a certain type of jaw suspension other than what modern sharks had, it didn't mean Troll had to change his fundamental vision of what the beast looked like. And after all, Pradel wrote, *everything is possible with Paleozoic fishes!*

The team scrambled for travel money, checking their own and other institutional couches for loose change. To stretch budgets, Pradel could sleep at Tapanila's house, and Wilga, Ramsay, and Troll would stay at the seriously budget-priced Thunderbird Motel across from campus. Flights were booked.

Pocatello is in the semiarid southeast corner of Idaho, with long, cold winters and hot, dry summers. The small city sits on the edge of the broad Snake River Plain at an elevation of 4,462 feet, surrounded on three sides by low, largely treeless mountains. In Western towns like this it's possible to see past the comfortably short buildings to the surrounding countryside. Pink light ran up the brown hills as Troll and Wilga, after meeting for the first time in person outside their doors at the Thunderbird, walked the six blocks to Elmer's Restaurant. For Troll, breakfast at Elmer's was one of the pleasures of trips to Pocatello. The morning sun comes in strong through the plateglass windows, the air is filled with the scent of hash browns, and career waitresses take your order and keep your coffee topped off. Troll and Wilga finished their eggs and, filled with a sense of excitement, walked across the street to campus.

13

TO THE SUMMIT AND BEYOND

It would be possible to describe everything scientifically, but it would make no sense; it would be without meaning, as if you described a Beethoven symphony as a variation of wave pressure.
—Albert Einstein (1879-1955)

IDAHO STATE UNIVERSITY HAD AN ENROLLMENT OF ABOUT 14,300 STUDENTS. THE campus conveyed a spacious feel, with buildings of ocher-toned bricks built in an Army Corps of Engineers style. ISU motto: The truth will set you free.

Troll and Wilga walked past the near life-size flat metal sculpture of a prehistoric bison that served as the signpost for the Idaho Museum of Natural History, and passed through a nondescript side entrance. The hall was quiet as they made their way to a door with a keypad entrance, where Pruitt let them into the Idaho Virtualization Lab. The lab, a research unit of the museum, occupied a single large, long room and had the friendly feel and décor of a student computer club run by underground artists. It was also one of the best, and one of the few, museum virtualization labs in the United States. The Smithsonian had one, as did a handful of other institutions, though none as interesting or as under the radar as this state-of-the-art facility in Pocatello, Idaho. The IVL's mission was to create a digital archive of rare artifacts from their own or any other museum's collections, and share those digitized artifacts as widely as possible with researchers, students, or curious amateurs.

Pushed up against a wall of large windows, the three main work stations—big double monitors where Pruitt, Schlader, and their colleague Nicholas Clement sat—were cobbled of abutting miscellaneous tables and desks drooping with cables. Work space for whatever interns might be around was catch-as-catch-can. A laser scanner, looking like a mutant telescope, stood on tripod feet ready to surface scan the project of the day, and gray and beige lockers of varying height and shape rounded out the furniture. The IVL was a few steps away from the museum exhibits area, and one floor above the collections.

Electronic file-sharing, email, and virtual meetings can facilitate scientific work in important ways, but there is nothing like the powerful synergy of people together in the same physical space, focused on a common goal, with nothing else interfering. They had gotten to know one another long-distance over the preceding months, and now Team Helico—Tapanila, Pruitt, Troll, Pradel, Wilga, Ramsay, and Schlader, but missing Didier—was finally convened under one roof. Each one brought something unique to the work at hand.

In the gossamer realm of the fates, Pruitt had brought scientific interest in *Helicoprion* back to life through his curiosity and perseverance. To the summit, he brought a tooth-by-tooth, volution-by-volution familiarity with the individual fossils, and a similar close-focus intimacy with the CT scans of the jaw material. He also brought a mechanic's ability to put broken parts back together.

Tapanila brought the auspices of Idaho State University, rolled together with Idaho no. 4 and the world's largest collection of *Helicoprion* specimens, not to mention the computing resources of the IVL. Adding to that, he brought a broad and organized scientific perspective of the historical and current *Helicoprion* research, and a growing comprehension of the *Helicoprion* genus gained through his and Pruitt's work to sort out the species.

Pradel brought insights into the hazy paleontological world of fossil sharks, as well as solid grounding in the shifting sands of Paleozoic chondrichthyan phylogeny. And set into his classically sculpted Gallic face, under a mop of dark hair, were some of the best eyes in the business for discerning details in CT scans of fossil fish.

Wilga brought her mastery of shark anatomy and feeding behaviors, keen perceptions and ideas, and tenacious problem-solving energy.

Ramsay brought his feel for the relationship of shape and function, his instinct for biological reverse engineering, and his formulas for calculating bite forces.

Schlader brought NASCAR-level driving abilities to the computer modeling software, and as an outsider to both biology and paleontology, a different vantage point and set of questions.

Troll brought twenty-plus years of *Helicoprion* passion and his sketchbook. Most remarkably, he had brought this diverse, unexpected, perfect group together.

The overarching goal of the two-day Shark Summit was for team members to put their minds and respective skills together to interpret the CT scan information, with a target of capturing their discoveries, insights, and conclusions in the first draft of a paper by the end of their time together. It would be like mathematicians or physicists working together at a white board, each one bringing a unique piece of the puzzle to solve an equation—questioning, tweaking, adding, subtracting, until everything falls into place.

Going in, they knew they had an upper and a lower jaw with one whorl between them. But how did it all fit together? How did it work? Were they missing a hyoid somewhere? The first task was to take a close look at the scans, computer model, and fossils and come to agreement on what they were seeing. The second task was to determine if they could glean enough information from the scans, cross-checking their observations and interpretations with one another, to confidently identify what type of jaw suspension *Helicoprion* had. If they could do that, they would be able to shed light on the important question of where *Helicoprion* fit in the long-surviving, multi-branching, ever-evolving class Chondrichthyes. Figuring out the jaw suspension would give them a leg up on the third task, parsing out how the jaw and whorl functioned. While not a task, the fourth important undertaking was to keep their minds and eyes open. They had brand-new evidence to sort out, interpret, and hopefully, ultimately, present to the scientific community.

To get oriented, the group spent some time in the IVL, where Schlader and Pruitt spun the model in computer space to reveal every possible view. The models were cool, but they were models—and once more in this long and winding *Helicoprion* story, curious people made their way down echoing flights of stairs to a museum basement to commune with the wonders of a lost world. On their way into the inner collections area, they passed the orange flatbed dolly where IMNH 36710 was resting, the giant *Helicoprion* fossil that first caught Pruitt's attention and had led to this day.

Inside the inner collections room, Tapanila set three of the best specimens out on a table covered with clean sheets of thin foam cushioning wrap, and he and Pruitt explained the significant features of each one. On Idaho no. 4, they pointed out the juvenile tooth arch, the grainy texture of preserved jaw cartilage, and the thin lines on the broken edge of the fossil that indicated buried cartilage.

As if that weren't spectacular enough, Tapanila brought out a small, flat box holding the rarest-of-rare *Helicoprion* relics: a loose splay of four teeth that had apparently popped out of Idaho no. 1 when Bendix-Almgreen was studying that specimen. Idaho no. 1 was one of the fossils Walter Youngquist collected from the shut-down Waterloo mine back in 1949. Although Youngquist didn't take both part and counterpart of all the fossils he whacked open that hot July day, he did put both halves of Idaho no. 1 in his car. Perhaps as Bendix-Almgreen was handling the specimen well over a decade later, the four tooth crowns popped out of the rock like a muffin from a tin. The original layer of enameloid was gone, but the inch-plus cutting blades were intact. These were the teeth, the real teeth, of a real *Helicoprion*, that may have swum the Phosphoria Sea that once ebbed and flowed right where they were standing. Using the two halves of the fossil as a mold, Bendix-Almgreen (with help from his conservator wife) created a cast of the teeth in PVC plastic. Although the actual piece that popped out only had four tooth crowns, Bendix-Almgreen was able to capture seven in his cast. One by one the Team Helico members peered closely at the extraordinary fossil, and turned the cast over and over in their hands, running their fingers over the pointed tips and scalloped serrations.

Energized, they took Idaho no. 4 into the "dirty prep" room, which had a tall table with what looked like a sandbox for its top. ("Dirty prep" rooms are like garage shops, while "clean prep" rooms are more like office space.) Tapanila carefully set Idaho no. 4 into the sandbox and they jostled around the table, talking at once, pointing out various features, correlating parts of the fossil with the scans, and tossing out questions and ideas. *So is this the back of the palatoquadrate? How thick was the Meckel's? What's this?* Thoughts and ideas ricocheted until the room was awash: *yes, but . . . what if . . . could be . . . unless . . . if that . . . how about . . . maybe . . . do you think . . . this could be . . . that makes sense . . . I'm not sure . . . if it was . . . I don't know . . . maybe so.* Troll circled and flittered, capturing the event on his iPhone like a proud father.

For the rest of that first summit day the group flowed like mercury, sometimes coalesced into a group, other times split up in singles or twos and threes to consider or discuss some aspect of the puzzle—sitting at a computer screen in the IVL, confabbing in the tiny lunchroom just off the lab, trooping downstairs to the collections or prep rooms to check something, sketching versions of *Helicoprion*'s head, or off in corners pounding out notes on laptops.

Pradel could only stay in Idaho one day, so at the end of that very full first day, he and Tapanila sat together in the little lunchroom and collaboratively typed up the first draft of what would be the description section of the team's first paper. Pradel was a native French speaker, Tapanila native Canadian, but they shared the language of paleontology. That evening Team Helico hit Pocatello's brew pub to celebrate. Troll made Pradel repeat the word *Helicoprion* into Troll's phone recorder, reveling in the way it sounded with a French accent, "el-eee-coh-[guttural]PREE-ohn." To Troll, it was "hell-ih-co-PRY-on," while Tapanila pronounced it "HEEL-ih-co-PRY-on," and Pruitt said "HILL-ih-co-PRY-on." Troll worked his table of shark comrades hard, trying to get them to spin out speculative gory details about the way his killer beast might have slashed and ripped, wanting blood and guts worthy of Shark Week, but his assembled scientists teased and joked in the warmest and most conspiratorial way, ordered more beers, and offered up fake theories that *Helicoprion* was a kelp eater. Troll was in heaven.

The next day, Tapanila and Pruitt took Troll, Wilga, and Ramsay out to the phosphate mine to meet engineer Dave George. The Idaho contingent knew the lean, gray-bearded George well—he was a fossil buff, and in his decades with the mine had collected an impressive number of *Helicoprion* and other fossils, sometimes by happenstance, sometimes by scanning the mine's reject concretion piles in his spare moments. George had put his fossils out on the table to show his visitors: *Helicoprion* whorls, fish scale impressions, and random remains. Then with special permission, the group went out to the active pit. The wind whipped under their hard hats as, nearby, equipment operators drove giant dozers, scraping layers of material off the pit wall to carry off for processing into fertilizer. The layers had been tilted by tectonic processes, but each line, each layer visible in the wall represented the Phosphoria seafloor over time. There had to be dead sharks in there. The group made its way to a rubble pile, where Pruitt demonstrated his concretion whacker tool, nicknamed the Persuader. The tool worked something like a fence-post driver, and was basically a sliding sledgehammer with a chisel on one end and a weight on the other. No one found a *Helicoprion* that day, but Ramsay did spot a brachiopod, his first fossil find.

Back on campus, they finished off the summit with a final visit to the *Helicoprion* fossils downstairs, and more time in front of Pruitt's computer to scrutinize the model with the new insights they'd gained. Tapanila and Wilga worked together drafting more sections of the paper as he and Pradel had done the day before.

For their last evening in Pocatello, Tapanila and his wife, Lori, a geologist specializing in metamorphic rock, hosted the group for a shark family dinner. As he had the night before, Troll circled the table with his iPhone like a documentarian, egging the group on to memorialize the monster shark and summarize what they'd learned over the past couple of days.

———————

The first paper appeared in *Biology Letters* in February 2013: "Jaws for a spiral-tooth whorl: CT images reveal novel adaptation and phylogeny in fossil *Helicoprion*." Tapanila was the lead author, with coauthors Pruitt, Pradel, Wilga, Ramsay, Schlader, and Didier. Because Troll wasn't associated with a university or museum he wasn't an official coauthor, but he provided artwork for the paper— the *Helicoprion* portrait that emerged from the summit, plus a school of "misbegotten helicos" illustrating all the versions that had come before, from Karpinsky's snout whorl to the Smithsonian's throat whorl. Troll's illustration of *Helicoprion* had gill slits, and was accompanied with the note, "configuration of gill slits and fins based on related fish, e.g., *Caseodus* and *Ornithoprion*."

The mainstream media as well as science bloggers jumped on the story, showing enthusiastic interest in the team's work to shed new light on this old mystery, and thrusting *Helicoprion* back into the public spotlight for the first time since F. John's cocoa cards and the *Popular Science* article at the dawn of the twentieth century.

Tapanila and Pruitt's species paper appeared next. "Unraveling Species Concepts for the *Helicoprion* Tooth Whorl" was the cover article in the *Journal of Paleontology* in the fall of 2013. They were the sole authors, but among the acknowledgments they thanked John Long and Ray Troll for valuable discussions.

The third team paper appeared in the *Journal of Morphology* in the fall of 2014, "Eating with a Saw for a Jaw: Functional Morphology of the Jaws and Tooth-Whorl in *Helicoprion davisii*." Ramsay and Wilga were the lead authors on that one, with the team thanking Troll in the acknowledgments, and giving an appreciative nod to Oleg Lebedev for his insights and suggestions. It was the top downloaded paper for that journal in 2015.

So just what did this bright assembly of mavericks and outsiders uncover about *Helicoprion*? What did they conclude about the Permian's biggest, weirdest shark? Let's start with the species work, then sink our teeth into the physical details.

SPECIES

Based on their tedious but productive months of morphometric measuring-measuring-measuring, Tapanila and Pruitt cleaned up shop on *Helicoprion* species. Since 1899, when Karpinsky transferred *davisii* from the *Edestus* genus and named the *bessonowi* species, at least eight additional species had been named (one could say, "had been named willy-nilly"). Through their analysis of twenty-three specimens, which included seven of the ten *Helicoprion*-type specimens (the fossils on which the declaration of a new species had been based), Tapanila and Pruitt concluded there were actually only three valid species in the *Helicoprion* genus: *bessonowi*, *davisii*, and *ergassaminon*.

H. *bessonowi* was distinguished by closely spaced, narrow teeth with short cutting surfaces. Tapanila and Pruitt surmised that *H. bessonowi* was the first *Helicoprion* species to evolve because *bessonowi* fossils are found in the oldest rock layers in the *Helicoprion*-bearing strata. *Bessonowi* fossils had been found primarily in Russia, and the type fossil was one of the specimens A. Bessonov sent to Alexander Karpinsky from Krasnoufimsk. That type specimen was at the center of an international crime story, when in 1998, it was found to have been stolen from its vault at the Natural History Museum in Saint Petersburg, Russia. John Long tells the story in his book *The Dinosaur Dealers,* which details the secretive world of international fossil poaching and smuggling. Fortunately, and with Long's help, the *bessonowi* type fossil was recovered and returned to its rightful place in Saint Petersburg after traveling through Finland, Florida, California, Paris, and London.

H. davisii had broad, widely spaced teeth with tall cutting surfaces, and appeared to be the most common and globally widespread *Helicoprion* species. Idaho no. 4 was a *davisii,* and it was the dominant species found in the Phosphoria. The first *Helicoprion* ever found, the fan of fourteen teeth that Mr. Davis picked up in the Australian outback, was the type fossil for *H. davisii.* It was in the collection of the Western Australian Museum in Perth, Australia.

H. ergassaminon had closely spaced, narrow teeth with tall cutting surfaces and was the least common species found. Bendix-Almgreen named *ergassaminon* in his 1966 *Helicoprion* paper, after noting Idaho no. 5 had unusual breaks and grooves on its teeth, which he interpreted as biting patterns. In a reference to the wear on the teeth, the name has its root in the prefix *erg,* meaning "work." The type specimen, Idaho no. 5, belonged to the University of Idaho, but the fossil is now missing.

In a clattering of calipers, it was out with the old. Gone from the taxonomic rolls were Hay's *H. ferrieri* (absorbed primarily into *davisii*), and Wheeler's *H. sierrensis* (*davisii* again) and *H. nevadensis* (reassigned to *bessonowi*). Gone were *H. jingmenense* (now *davisii*) and *H. mexicanus* (marked *nomen dubium*, or "doubtful name," to reject the ten-tooth-crown specimen as its own species, but acknowleding insufficient information to definitively place it). Lebedev said the paper was "a beautiful study and it was long ago high time to revise the systematic composition of the genus."

WHORL PLACEMENT

The CT scans of Idaho no. 4 clearly showed that the whorl filled almost the entire span of the lower jaw. Unlike what Bendix-Almgreen and Lebedev had thought, the whorl was not housed at the end of an elongated lower jaw. Rather, it filled the jaw. After more than a century, this most persistent, fundamental question could be put to rest.

JAW SUSPENSION AND PHYLOGENY

At the Shark Summit, the elephant in the room—no longer surrounded by blind men but by astute and specialized observers with their eyes wide-open—was jaw suspension. Could they ascertain and confirm how *Helicoprion*'s jaw was suspended from its cranium? The CT scans captured the backs of the jaws where the jaw would have been jointed, and also imaged the entire upper jaw except for the very front part. That should give them enough to go on.

As the team expert in the evolutionary lineages and relationships of chondrichthyans, Pradel best understood the morphological clues to look for. Even before the summit, he had said that from what he knew about the eugeneodontids, he would expect *Helicoprion* to exhibit an autodiastylic style of jaw suspension. As we've learned, hyostyly is the flexible style of jaw suspension, supported by the hyoid arch, favored by modern sharks. In contrast, the fused and rigid holostylic jaw suspension is found in modern holocephalans. The autodiastylic suspension Pradel referenced was a sort of intermediate system, thought by many researchers to be the possible ancestral jaw suspension type for all chondrichthyans. Autodiastyly is not found in any living chondrichthyans, although it could be seen in a developmental stage of holocephalan embryos. (Embryology allows scientists to observe aspects of evolution in fast forward, like how human embryos begin with gills but develop lungs before birth.)

In autodiastyly, the palatoquadrate was suspended from the cranium in two places: in the ethmoid region around the nose, and farther back, by the otic—or ear, or basal—region. As Toby White explains it on palaeos.com, "From this version of autodiastyly it is straightforward to derive both holo- and hyostyly, as well as other variations on those themes. Either the palatoquadrate fuses with the base of the braincase for support as it grows (holostyly), or it enlists the aid of other structures such as the otic capsule or the hyomandibular (autostyly, hyostyly)."

The autodiastyly story began with a British evolutionary embryologist and paleontological enthusiast named Gavin Rylands de Beer. In 1935, de Beer and British paleoicthyologist James Alan Moy-Thomas predicted the existence of autodiastyly, in which the hyoid arch wouldn't be involved in jaw suspension, but instead, the upper jaw would be strongly suspended by ligaments from the cranium at two points, in the ethmoid and otic regions. Some sixty years after de Beers and Moy-Thomas predicted autodiastyly, Eileen Grogan and Dick Lund reported they had found the first real evidence of it, in a 320-million-year-old Bear Gulch chondrichthyan they named *Debeerius ellefseni*. *Debeerius* was a foot-long, chimaerid-looking chondrichthyan, with a big head and delicate, wing-like pectoral fins. The fossil even preserved skin pigments, which revealed a tiger striping. Lund and Grogan used *Debeerius* to support their argument that class Chondrichthyes should be divided into the subclasses of Elasmobranchii and Euchondrocephali. Lund and Grogan included eugeneodontids among the Euchondrocephali.

Again drawing on White's inimitable explanation:

> *The issue goes back to the Hundred-Years' War over the ancestry of holocephalan relationships. . . . Few people (Bashford Dean, perhaps, excepted) study ratfish for their intrinsic beauty, for spiritual inspiration, or for the millions undoubtedly to be made peering at dried tritors. No, it is only the jaw that matters. Chimaeroids are the most primitive of living gnathostomes [jawed vertebrates]. To find the plesiomorphic [ancestral] state of the holocephalan jaw is, just possibly, to find the primordial state of all jaws. And, from there, it is but one step to the Holy Grail of vertebrate paleontology—the origin of jaws.*
>
> *A few years ago, the intrepid knightly band of Eileen Grogan, Richard Lund and Dominique Didier (Grogan et al., 1999) may, like Lancelot, have partially completed the Great Quest for this Grail, or at least have seen a vision of it, "All palled in crimson samite, and around/Great angels, awful shapes, and wings and eyes." Well—perhaps not the bit about the angels and wings. But*

the eyes and, definitely, most definitely, the awful shapes. The
included smeary photomicrographs are certainly that. . . . They
found that the primitive state for all chondrichthyans (and quite
likely placoderms as well) is autodiastyly.

Pradel didn't completely agree that autodiastyly was the characteristic primitive state of gnathostomes—although in the end, he
granted, who knew? In any case, he predicted autodiastyly for *Heli-*
coprion. But what would they find?

Team Helico peered at the CT scans. Pradel pointed out the
shape of the palatoquadrate, which was similar to that of *Debeerius*
and very different from the palatoquadrate of modern sharks. The
shape was also notably different from the typical palatoquadrate
of a common group of Paleozoic sharks known as "cladodonts,"
which included the very ancient *Cladoselache* as well as the
"ironing board" shark, *Stethacanthus.* (The word "cladodont"—
cladus, meaning "branch," and *odon,* meaning "tooth"—referred
to the group's shared tooth style, which featured some variation
of one central spike or blade surrounded by multiple shorter
prongs.) The cladodont palatoquadrate had a long process, or projection, on the upper side at the back of the orbit, or eye area. That
cladodont-type process was known to not be present in euchondrocephalans—now including, the team determined, *Helicoprion,*
as represented by Idaho no. 4.

On the *Helicoprion* palatoquadrate, Wilga identified two possible
attachment points, ethmoid and otic, that would be characteristic
of autodiastyly. Clearly, the palatoquadrate in the scans connected
with the braincase at the key ethmoid and otic locations. Idaho
no. 4 had ample intact rock above the palatoquadrate, yet the scans

didn't pick up any parts of a neurocranium. If the palatoquadrate had been fused, as in more modern holocephalans, that would have been evident in the scans. The scientists in the group agreed: *Helicoprion* had the autodiastylic jaw suspension of euchondrocephalans.

Between Pradel's strong French accent and the confusing language of phylogeny, all Troll heard was ". . . cephalan," which to him meant only one anguishing thing, his beautiful monster shark was a ratfish. That's not what it really meant, but as an artist with a two-decade investment in the sharky look of *Helicoprion*, and an equally long history of drawing rabbit-faced, so-ugly-they're-beautiful ratfish, that's how he interpreted it. *You're saying Helicoprion was a ratfish?!*

No, not a ratfish. Ratfish are modern. *Helicoprion* wasn't even technically a ratfish ancestor, because the buzz saw shark was an evolutionary dead end. To help clarify, Pradel sketched out a theoretical cladogram on the whiteboard in the museum's tiny break room. (Cladograms have replaced family trees for illustrating evolutionary relationships, and look more like bushy shrubs than trees.) Yes, *Helicoprion* was among the chondrichthyans from which certain subsets, some day, would emerge as holocephalans. No, *Helicoprion* was not in the bush headed for modern sharkhood. But that was almost beside the point in terms of talking Troll back from the ledge. The most important point Pradel was trying to make was that the Paleozoic was a time of extreme chondrichthyan diversity. Modern sharks and modern ratfish both had autodiastylic ancestors. So *Helicoprion* could indeed still look like a shark.

The scientists threw life rings to Troll. *Helicoprion*'s teeth were definitely sharklike, Pruitt offered. *Helicoprion* and great white shark teeth looked the same in cross section, whereas a cross section of a holocephalan tooth look different. So *Helicoprion* did have teeth like a shark—although yes, it was true, the whorl did grow symphysealy, which is a holocephalan trait. *It's really weird, but it's a shark,* Pruitt said. *It's fair to call it a shark,* Pradel agreed. Troll would not have to change the look of his *Helicoprion.*

A casual observer would be tempted to ignore the whole brouhaha as a matter of semantics. Shark, sort-of-shark, whatever. But this weirdness—sharklike teeth on a symphyseal structure—was a very specific evolutionary declaration of independence in the eugeneodontid tribe. Yes it was weirdness, but it was their very own, very novel weirdness. This is exactly the weirdness that gave scientists such fits, and was the reason it took them so long to figure these animals out. Miss Fanny Hitchcock had the initial glimmer of understanding and was the first to try to explain it, in her paper proposing that *Edestus* blades were teeth, not fin spines. The confusing twist was that *Edestus* teeth weren't imbedded in a jaw like normal vertebrate teeth, but were what she called "intermandibular" teeth, supported by some sort of structure *between* the jaws. This is what set the eugeneodontids apart, as well as a few other groups, including those phylogenetic gypsies, the iniopterygians.

As long as *Helicoprion* could look like a shark it could have plaid suspenders, as far as Troll was concerned. He was satisfied.

In the *Biology Letters* paper, the team wrote, "CT scans demonstrate that *Helicoprion* possessed an autodiastylic jaw suspension characterized by a two-point articulation of the upper jaw to the neurocranium via ethmoid and basal processes. . . . An autodiastylic jaw suspension is diagnostic of euchondrocephalans, which confirms . . . *Helicoprion* [belongs] among the Euchondrocephali. This result provides new insight into the evolutionary history of early holocephalans, including their high degree of specialization and large body size during the Late Paleozoic, which may correspond to the increased diversity and abundance of cephalopod prey at this time."

It was a highly significant declaration, the finer points of which were lost on, or confused by, anyone not well-versed in the phylogeny and classification of Paleozoic sharks—which meant just about everyone reading (or writing) a newspaper or blog, or watching (or narrating) television and radio reports on *Helicoprion* after the news broke in 2013.

Author Brian Switek, who had written about *Helicoprion* following Lebedev's paper and who had corresponded with Troll, knew more than most journalists about the buzz saw shark. Switek was among the first to report on the team's breakthrough findings, and for the most part he wrote with scientific exactitude and historical context. He did make one understandable but erroneous leap however when he described *Helicoprion* in his blog as having the characteristics "of a group of cartilaginous fish called Euchondrocephali—commonly known as ratfish and chimaeras."

In the comments section of Switek's blog, edestid researcher Wayne Itano wrote, "My reading of the journal article is that they are claiming *Helicoprion* is a Euchondrocephalan and stem-group holocephalan, not crown-group holocephalan. So calling *Helicoprion* a ratfish rather than some sort of a distant relation of a ratfish doesn't seem quite right." John Maisey expressed a similar sentiment.

It's hard to write about science without raising at least a few academic eyebrows. And it's no surprise that instead of taking the mind-numbing plunge into chondrichthyan phylogeny, reporters in the mainstream media leaped far more gleefully straight to ratfish. In the news, *Helicoprion* became a giant freaky ratfish.

And there we have the greatest sticking point when talking about *Helicoprion*. What do you call it? Most correctly you call it a euchondrocephalan, but no one will know what you're talking about. To call it a ratfish is fun and gets people's attention, but it's not accurate. If you call it a shark, you will get pushback from the scholars. So what to do? Introduce *Helicoprion* to interested but uninitiated listeners as a shadowy harbinger of modern sharks, to conjure the right sleekly swimming visual image. Then tell them they might be surprised and fascinated to learn that this commanding ancient predator was on a fundamentally different ancestral branch than today's "true" sharks. Explain that even though *Helicoprion* was at the top of the marine food chain for millions of years, it nevertheless went extinct under uncertain circumstances, at the dead end of an evolutionary line, leaving no direct heirs. *Helicoprion,* with its extraordinary tooth whorl experiment, was literally one of a kind. If your listeners are ready for more, lean in and reveal that this monster beast with the buzz saw teeth was actually more closely related to today's ratfish than to today's great whites. While they're processing that, reference a "lost world" to set the hook and make them hungry for details.

LABIAL CARTILAGE

One of the most puzzling aspects of *Helicoprion* was how the tooth whorl was mechanically affixed in the mouth. Normal shark teeth are imbedded in tissue anchoring them to the jaw. This is an unapologetically temporary arrangement, given the way shark teeth continually grow and shed—but the way they are moored while doing that toothy work makes logical sense. *Helicoprion*'s single tooth, however, wasn't bound to the jaw in any way. It was imbedded *between* the right and left halves of the lower jaw, positioned at the symphysis, or center front. But was it just jammed there like a pizza cutter pushed halfway into a quart of ice cream? How did it stay centered as it built its successive volutions? What happened as the whorl added new crowns and older ones disappeared into the jaw?

Team Helico agreed with Bendix-Almgreen's elegant conclusion that the whorl reeled itself in via a geometry initiated by the juvenile tooth arch. The team further articulated that as older teeth were resorbed into the inner sanctum of the lower jaw, those buried parts became covered by a tessellated, encapsulating cartilage that entombed the buried tooth crowns and hard, dentinal root-base. Remnants of that encapsulating cartilage were visible on the surface of Idaho no. 4. The encapsulating cartilage likely formed from the same physiological responses that enclose foreign objects in scar tissue to shield surrounding soft tissues from injury. But would that imbeddedness and encapsulation provide a strong enough structure to keep the whorl from wobbling when *Helicoprion* bit down on struggling prey, or from wandering off center as it grew?

One of the advantages of "digital prep," as Pruitt liked to call it, was that he could cut and paste parts—that is, outline well-defined shapes visible in the scans, digitally cut them out, and move them around. It was essentially the same thing fossil preparators did with "analog" fossils—pick up bone-part A and see if it fits onto bone-part B. Animals often become disarticulated (disassembled) before they are fossilized, disturbed by moving water or scavengers, for instance, so the pieces need to be reassembled. Pruitt had noticed three prominent pieces in the scans that seemed to be broken parts of the same piece of cartilage. When he digitally mended the parts, the resulting cartilage took roughly the shape of a slice of pizza. Where did it fit? What was its function?

On the second day of the Shark Summit, Pruitt, Wilga, and Tapanila were clustered around Pruitt's work station discussing jaw joints while Pruitt was messing around with the mystery cartilage. He rotated the piece on his screen for a better look and noticed a sort of ledge on the broader end of the piece and a slightly dished, faintly stair-stepped inner edge on the other end. Pruitt shifted the orientation of the piece a little bit and lo and behold, the ledge slotted into the lower jaw. The other end fit neatly at the whorl's outer volution. It suddenly looked less like a slice of pizza and more like a strut. *Hey, guys, check this out!*

Modern chondrichthyans have specialized "labial cartilages" that support the corners of the mouth. (*Labial* meaning "of or related to the lips.") Typically, labial cartilages pull the skin forward when the fish opens its mouth. If you've set up a modern tent lately, you can picture the flexible tent poles that snap the tent fabric tautly into place. The piece Pruitt was fiddling with was in the mouth region all right, but it didn't look like any labials Wilga had ever seen, and she had about seen them all. In modern fish, the action of the labials makes a tunnel of the mouth and also helps create a vacuum for suction feeding. It wouldn't make sense for *Helicoprion* to have been primarily a suction feeder, because it didn't have a robust hyoid arch to create that bellows-style vacuum, and suction feeders wouldn't have a big whorl of cutting teeth blocking the way to the gullet. *Helicoprion* had the teeth of a bite feeder, and wouldn't have wanted a screen of skin blocking the whorl from the sides.

The team wasn't immediately sure what it was, so in the days following the summit, they debated long-distance if this strut-like cartilage might be a growth, or "process" (think noun, not verb), from the lower jaw (or "Meckel's cartilage"). Wilga especially rejected that possibility in favor of the piece being a modified labial cartilage. Why would evolution build an energy-expensive new extension like that when it already had all those nice labial cartilages to work with like so much biological Play-Doh? As we've seen with jaws, evolution likes to repurpose existing parts, even down to the molecular level, rather than create something from scratch. Plus, if this were a new extension or giant Meckel's process, there wouldn't be a joint at the jawline, and the scans clearly showed a joint, not a fracture line, where the strut met the lower jaw. (A joint is where two parts come together. There are hinged joints, like your knee, and there are junction-joints of two cartilaginous elements, called "synchondrosis." The joint where the labial cartilage met the lower jaw was a synchondrosis.)

Ultimately, Pradel and Wilga agreed the structure could reasonably, and most conservatively, be identified as a labial cartilage—even though this was a novel use of labials and an unprecedented structure, never seen in any other chondrichthyan. As the team reported in *Biology Letters*, this labial cartilage must be what supported the whorl and what kept the root spiral aligned as it grew. This buttressing, guiding labial cartilage was as unique to *Helicoprion* as the tooth whorl—and is apparently what enabled the success of the whorl. The labial was a novel evolutionary adaptation to a novel evolutionary adaptation.

The lower part of the labial strut cupped under the bottom of the whorl, holding the base of the root inside the lower jaw. The upper part of the strut rode against the top part of the whorl. Based on her experience, Wilga proposed a gliding joint for that point of contact. Gliding joints can occur where two relatively flat surfaces meet, and allow those surfaces to move in any planar direction—like stacking two hockey pucks and slipping them around on each other. Modern sharks have gliding joints where the palatoquadrate and cranium meet, and we humans have gliding joints scattered in strategic places from our necks to our feet.

After the *Biology Letters* paper had already been published, Pruitt and Tapanila had a high-wattage light bulb moment when they realized a smooth, slightly dished impression pressed into IMNH 29094—a big three-and-a-half-volution tooth whorl from the Gay Mine—was actually an impression of the labial cartilage. They had noticed the impression early on in their work with the fossils, and had even talked about it, ultimately dismissing it as something that may have pressed against the whorl during fossilization. Which is exactly what it was. Pruitt suddenly recognized it when he was digitizing the surface of the fossil, which takes all the color out, revealing subtleties of shape. He flipped the shape around digitally so he could see it popped out instead of pressed in, and ran for Tapanila. What they had was an impression of a labial cartilage riding perfectly in place on the whorl. The slight stair-stepped cupping of the top end of the labial, as revealed in the CT scans, matched the volutions of the root. Nature always has a plan, even when it doesn't know where it's going.

PRUITT'S PROCESS

Noodling with the labial cartilage on the computer led to the next big reveal—how *Helicoprion* didn't slice through its own brain when it chomped down on something.

Tapanila and Wilga were sitting at Pruitt's work station discussing jaw joints, including the question of what stopped the jaws in closure since there were no upper and lower teeth to occlude. There was a cavity to accept the whorl when the shark closed its mouth, but what prevented the whorl from "over-closing," and piercing the upper palate?

Pruitt had put the labial cartilage in place as a strut, and was listening to the conversation with one ear as he toyed with a small piece of cartilage floating unattached in the scans. It looked like a piece of something, not just fossil flotsam. Pruitt was trying hard to make the piece fit somewhere into the labial cartilage, flipping it around every which way, but nothing was lining up. Frustrated, he dragged the piece over to the back of the jaw, near the joint that Tapanila and Wilga were talking about—and the piece virtually clicked into place, as a peg, or a knob (a "process," in the lingo), on the lower jaw that lined up perfectly with another peg-like projection on the upper jaw. The two would butt against each other to act as a stop when the jaw joint closed, while allowing the roof of the *Helicoprion*'s mouth that necessary level of clearance above the whorl.

Once Pruitt had this model in action, the team gathered around Pruitt's screen, making him move the piece in and out of place until they also were convinced it was real. Most of the discoveries and understandings the team came to during the summit were gradual or cumulative—but finding the answer to the jaw-stopping question was a eureka moment. The obvious code name for this important mechanical feature had to be the Pruitt Process, even if the vocabulary word in the papers would be posterior process.

By the time the team published their "Eating With a Saw for a Jaw" paper in the *Journal of Morphology*, they had determined that the labial cartilage also helped transfer the jaw's closing force over a larger area, taking some of the pressure off the Pruitt Process, and working in tandem with it to stop the jaw from over-closing.

JAW MECHANICS

During the Shark Summit, Ramsay frequently sequestered himself to work on the challenge of estimating *Helicoprion*'s bite force, first by trying to break down the mechanics of how the jaw operated so he could represent the information numerically. He started by sketching what he thought the musculature would be, based on the muscle attachment points they had identified.

Troll was especially interested in Ramsay's activity—for one thing, because Ramsay was also an artist, and also because more than any of the others, Ramsay's questions directly related to how the shark's head might look physically, and how it would have used its tooth whorl, the focal point for the beast in all its glory. The two artists traded notions as the picture filled in over the course of the summit.

Ramsay was concentrating hard on how to represent the bite force data graphically, visualizing vectors and levers, scribbling formulas, and moving toward a calculation for mechanical advantage. Mechanical advantage is a common biomechanical measure used to determine how the force and velocity that muscles generate is transferred through skeletal elements.

Pruitt and Schlader had used a 3-D printer to create a working model of the Idaho no. 4 jaws for the summit, complete with tooth whorl. Taking a break from the physics and math, Ramsay picked up the 3-D model and started playing with it. He was mesmerized by the way the teeth tracked as the jaws opened and closed. Continuing to open and close the jaws, he observed that the teeth traveled in an arc. Ramsay had just finished his PhD and had spent a lot of time thinking about teeth and how teeth cut. As part of that study, he read up on the mechanics of slashing weaponry. The whorl was going through an arc similar to a curved sword or knife.

Then the angles of the cutting edges of the tooth crowns came into focus for him. *Wait a minute.* Both the forward-facing and backward-facing edges of the tooth crowns had serrated cutting edges. This meant that the whorl sliced with its tooth crowns as the jaws closed over a prey item, and would have kept slashing with the opposite edge of the tooth as the jaws opened. *Helicoprion*'s teeth cut in two directions. The whorl really *was* a saw. A slashing saw! The cutting edges on the teeth of most modern bite-feeding sharks run basically parallel to the jaw. But *Helicoprion*'s teeth were perpendicular. That aha moment led to the next big idea. Many modern predatory sharks leverage the cutting action of their teeth by shaking their heads when they have prey clamped between their jaws. When they do that, they are in effect forcefully sawing whatever meat is held in their teeth. So the "processing" of prey by modern bite feeders is a two-step process: the shark clamps prey in its teeth to capture and puncture the body, then shakes its head from side to side to saw and cut. It looked to Ramsay like *Helicoprion* had combined two effective prey-processing behaviors, biting and shaking, into one.

There was more. Not only did *Helicoprion*'s teeth slash in two directions, the whorl had additional efficiencies, which became clear as Ramsay calculated out the details in the months following the summit. He used the muscle attachment points to infer the orientation and nature of the muscles, which helped him determine lever mechanics and mechanical advantage. The math and physics revealed that tooth function and force changed along the whorl as the jaws closed. In the eleven or so teeth exposed in *Helicoprion*'s mouth at any given time, the newer, larger teeth deeper in the mouth had greater mechanical advantage than the older, smaller teeth more forward in the mouth. Imagine a *Helicoprion* swimming toward a juicy Permian squid. It powers toward the squid with its mouth open at (according to the calculation) seventy-seven degrees, an angle that falls into the natural range of modern bite-feeding sharks, and which would have been wide enough to keep the whorl from blocking the gape. With a final thrust of its powerful tail, the *Helicoprion* grabs a good mouthful of the squid. The largest teeth toward the back of the whorl would have grabbed and punctured the squid at first contact, and as the mouth continued to close, those big teeth would have pulled the squid farther back into the mouth toward the shark's gullet. Also as the mouth was closing, the smaller teeth toward the front of the whorl were moving with a higher velocity than the larger teeth to snag any squid parts that might otherwise slip away, also pulling those bits into the mouth as the jaws closed. Snag and drag. Once the first fatal bite was made, the shark probably made smaller biting motions to hack the prey and push food farther down the gullet.

The bite forces that Ramsay calculated for *Helicoprion* were huge, and exceeded those of most modern fish and mammals. It seemed excessive for an animal that would have been eating soft-bodied prey, but bite force isn't just how hard the jaws can clamp down, it's also how fast. Consider the catfish, with its puffy cheeks. That puffiness comes from the catfish's big jaw muscles, which allow it to snap its mouth closed really, really fast. If *Helicoprion* were chasing a cephalopod with jet propulsion, it would have wanted to bite fast when it had the chance.

UPPER DENTITION

While the tooth whorl naturally gets all the attention, one of the persistent questions in *Helicoprion* research has been what sort of teeth, if any, were associated with the palatoquadrate? The work of Tapanila et al. sealed the deal that *Helicoprion* had only one tooth-whorl, situated in the lower jaw. But what was going on up top? Bendix-Almgreen wrote that the upper jaw held "small tooth-like elements," based on features he observed on the Idaho no. 4 fossil. He described those elements as nearly rectangular in shape and growing in rows. Lebedev also proposed that *Helicoprion* had small upper tooth elements. He had dissolved part of the matrix from around his whorl fragments and recovered microremains, some of which he identified as tiny brick-like teeth, and called lateral dentition for the way it lined the sides of the palatoquadrate. Lebedev likened the lateral dentition to "chain armor" that would provide semirigid support and serve as both gripper and cutting board to allow *Helicoprion* more effective use of its whorl.

In their *Biology Letters* paper, the team had no CT evidence to report of lateral dentition, but new information emerged later after Tapanila tracked down a fossil that Bendix-Almgreen included a photograph of, but had not discussed, in his paper. Tapanila found the revealing fossil at the Smithsonian—the so-called "Sweetwood specimen," collected from the Gay Mine and turned over to science by C. W. Sweetwood and "authorities of the Simplot Fertilizer Company." In supplementary material for the *Journal of Morphology* paper by Ramsay et al., Tapanila provided images and a full description of palatoquadrate teeth—decisive evidence of very real, very numerous pavement-style teeth. It was a good system. The peaked upper jaw, lateral dentition, and whorl would have worked in harmony to crimp prey. It looked like *Helicoprion*'s full predatory modus operandi was snag and drag, crimp, bite-bite-slash-saw-swallow.

WHAT WAS *HELICOPRION* EATING?

But what was the object of this snag-drag-crimp-bite-slash-saw-swallow sequence? The snagging, slicing shape of the tooth crowns, coupled with the fact that *Helicoprion* teeth rarely showed much obvious wear, indicated the shark was eating soft-bodied prey. Cephalopods and small chondrichthyans with soft cartilaginous skeletons were abundant in the early Permian seas, and made sense as *Helicoprion*'s prey of choice (as Lebedev had suggested). In fact, the abundance of cartilaginous fish and cephalopods may have had triggered *Helicoprion*'s specialized adaptation, the tooth whorl, to take advantage of those bountiful resources and grow into one of the largest and most successful predators of its time. Some Permian cephalopods didn't have external shells, but many, many did, like the ammonoids and nautiloids. What about those? *Helicoprion*'s tooth whorl was obviously not built for crushing shells, and as mentioned, evidence of strong wear marks on *Helicoprion* teeth was rare. Besides that, it would have been like trying to cut a walnut with scissors. The shell would pop right out.

While the team's "Eating with a Saw for a Jaw" paper was undergoing peer review, Ramsay received the comment from an anonymous reviewer who wrote something to the effect of: *Have you thought about this? What if* Helicoprion *caught the soft body part of an ammonoid and managed to shuck the animal out of its shell in the process?* As soon as Ramsay saw the comment, he could picture it. It made so much sense. If a *Helicoprion* swam up behind a fleeing ammonoid and bit it straight-on, the teeth toward the back of the whorl could snag and hold exposed parts of the head and arms, while the forward teeth caught flesh near the shell opening. As the shark's mouth closed, it might pinch the body out of the shell. At the very least it would have resulted in a big chunk of cephalopod, and at the very best, it would have shucked the animal. So add shuck to the sequence: snag, drag, shuck . . . Remarking on *Helicoprion* later, the eminent French paleontologist Philippe Janvier likened the tooth whorl to a *fourche d'escargot*, a snail fork.

THE PORTRAIT

During the Shark Summit, Troll was by turns thrilled and horrified, satisfied and frustrated. With each new step of understanding, he tweaked his sketches. As soon as he said he wouldn't draw the snout over again, he drew the snout over again. Throughout his long relationship with *Helicoprion*, Troll had the shark mostly to himself, and had been in the driver's seat as to how it looked. He had been free to pick and choose, based on his learned understanding of what the shark was, yet also on his sense of what he wanted it to be—even as he solicited, and genuinely wanted, input from anyone who might really know. Zangerl gave him evidence-based permission to illustrate *Helicoprion* as a giant shark, but then Zangerl also initially told Troll the beast had upper and lower whorls. Bendix-Almgreen's response to Troll's *Helicoprion* body reconstructions was that they could only be regarded as "pure conjecture." Lund and Grogan had instructed Troll to erase the gill slits. All John Maisey said of *Helicoprion* was that it swam, had a big tooth whorl, gills of some sort, calcified cartilage, and it ate stuff. That was all anyone could rightly surmise.

Yet despite scientific skepticism, there was a beast demanding to be seen, or at least glimpsed from imperfect clues. At the summit, Troll's singular, synthesized vision, assembled over more than twenty years, transformed into a shared vision. It was what he wanted, and what he had resisted. *You guys are taking my giant fish-shredding, prey-chopping monster beast and turning it into a squid eater!* Still, the truth remained that thanks to its ability to exploit a large prey base, *Helicoprion* became one of the largest predatory fish in the global oceans, cruising the top of the food chain for ten to fifteen million years. It was unequivocally the king of the realm, with the wildest whorl of slashing choppers the animal kingdom ever produced.

The team teased, cajoled, supported, humored, and listened to Troll throughout the summit. They were the scientists, but he was the archivist, the fan club president, the glue, the quarterback, and the one who had connected them all.

In the weeks after the summit, while the team was preparing the paper for publication, Troll continued to work with Ramsay and the others to capture and reflect the newest vision of *Helicoprion*. By the time the paper came out, he had created a series of half a dozen *Helicoprion* portraits with incremental shifts toward the final reconstruction. It was a beautiful beast.

14

SHARK IS A VERB

*It is the tension between creativity and skepticism that has produced
the stunning and unexpected findings of science.*
—Carl Sagan, from *Broca's Brain: Reflections
on the Romance of Science*, 1979

NOT SO FAST, SAID LUND AND GROGAN.

After the *Biology Letters* paper came out, paleontologists Dick Lund,
Eileen Grogan, and a smattering of others withheld full buy-in. Yes,
Helicoprion appeared to have had an autodiastylic jaw suspension. But
Grogan and Lund perceived too many functional improbabilities to
agree with the team's conclusion that the whorl extended along the
entire length of the lower jaw. They were concerned that the CT scans
might have reflected a preservational bias on that detail; in other words,
that the whorl and jaw had been shoved together into that position
sometime after the animal's death. Based on what they observed in
their Bear Gulch euchondrocephalans, they thought the whorl could
have been farther forward, perhaps on a labial cartilage at the front
of the mouth. If the whorl occupied the entire lower jaw, Lund and

Grogan wondered, what was going on in the rest of the head? Especially in terms of that nagging issue, the potential for the whorl to "disturb" the brain? They felt the topic wasn't adequately covered in the paper, and believed there might have been other ways to interpret the scans.

You probably know by now, as a close observer to this tantalizing saga of skill and chance, an uncontested win would have been too contrary to the great contrarian *Helicoprion* tradition. Fin spine! Teeth! Two whorls! One whorl! Outside the mouth! Inside the mouth!

Team Helico, through their rigorous work and committed, creative, unconventional collaboration, had achieved the first major scientific breakthrough in the understanding of *Helicoprion* in over a century. They documented and mapped the first in situ jaw material, they determined the nature of *Helicoprion*'s jaw suspension, and they authoritatively emended the species classifications, providing the community of paleo shark workers with a database and identification tools. As for the big question of exactly where the whorl fit on that peerless, unprecedented creature? Team Helico called the bet and put their cards on the table. *Okay then, prove us wrong.* And that's the nature of the beast. That's the way research works. The beautiful, frustrating, addictive, rewarding way it works.

As they had to, Grogan and Lund also took exception to Troll's illustrating *Helicoprion* with gill slits instead of a single opercular cover. Hadn't they explained that years ago when he visited them in Bear Gulch? Autodiastyly was not simply about jaws, they contended. You had to consider the hyoid. True, the hyoid wasn't involved in autodiastylic jaw suspension. But that's because in holocephalans, and at least some euchondrocephalans, the hyoid—that second repurposed gill arch—had taken on the job of supporting the operculum. As Toby White wrote on palaeos.com, "The chimaeran hyoid is, it turns out, quite happy supporting the operculum and has no interest in the palatoquadrate." In other words, the hyoid was giving structural support to the gill cover, and not the jaw. Grogan and Lund's autodiastylic *Debeerius* fossil showed "chimeroid-like gill baskets," which would have been covered by an operculum, a condition they proposed as widespread among the euchondrocephalans. In their minds, there was no scientifically based reason to reconstruct *Helicoprion* with gill slits, especially since the *Biology Letters* paper didn't include any evidence to support that interpretation. Lund and Grogan believed, as they had since 1998, that Troll should default to the opercular cover.

Dominique Didier tended to agree with Grogan, her longtime friend and colleague, on the operculum issue too. Didier felt the primary pro-gill-slit argument, that *Helicoprion* was a fast, open-ocean animal, was weak. Tunas are fast. Sailfish are big and fast. Wilga on the other hand countered that tuna and sailfish are bony fish—an entirely different clade. Apples and oranges! All the fastest swimming sharks and rays had large, long gill slits. The whole imbroglio over gill slits reminded Didier why she preferred biology to paleontology. Biologists can catch a fish and look at it. She didn't press the opercular point, as there were arguably decent arguments on both sides.

In addition to the knotty problem of how to illustrate *Helicoprion*'s gills, Troll faced another aesthetic choice in his portrayal of the whorl-toothed shark, the burning question of whether or not the beast had claspers. Claspers are the external inseminating organ of male chondrichthyans—sperm delivery units, though not technically penises. They come as a pair (although the male uses only one at a time), and lie inboard of the pelvic fins and flat along the underside of the body. For both extant and extinct species, claspers are one of the key features that define membership in class Chondrichthyes, in addition to the possession of a cartilaginous skeleton and placoid scales.

"All mature male chondrichthyans display intromittent [external copulatory] organs," say Grogan, Lund, and Emily Greenfest-Allen in their chapter on early chondrichthyans in *The Biology of Sharks and Their Relatives*. "The Upper Devonian *Cladoselache* has often been claimed to be the exception, yet such arguments ignore the high probability that the recovered forms are strictly female." Perhaps the fossil beds had once been shallower water zones where females went to pup, and not frequented by males. Also, we know that males and females of some modern sharks, including the mako, segregate by gender during parts of their life cycle, and that hammerheads, for instance, are known to aggregate in large female-only schools.

Rather incredibly, claspers actually can fossilize. In 2009, John Long, Per Ahlberg, Kate Trinajstic, and Zerina Johanson wrote about the discovery of claspers in a 380-million-year-old fossil placoderm. Rainer Zangerl, in his *Handbook of Paleoichthyology*, noted that eugeneodontid skeletons tended to be weakly calcified (therefore not ideal for fossilizing), but he also speculated to Troll that eugeneodontids didn't have claspers or anal fins—a proposition that had filtered through the paleo shark community as one way to explain their absence in the fossil record. Based on what he heard from Zangerl, Troll had never drawn claspers on his whorl-toothed sharks either. He did give megalodon a pretty good pair in *Sharkabet*, and sometime later, when someone who knew something about sharks asked Troll why he only illustrated female *helicoprions*,

Troll polled his sources on the matter. He first emailed John Long in Australia, a natural choice since Long had written a book on the prehistoric origins of sex, *Dawn of the Deed*. Long's response was that the notion of eugeneodontids lacking pelvic fins needed to be questioned and reinvestigated, because all chondrichthyans had claspers, no exception to the rule. And that would include *Helicoprion*. Long looped John Maisey and Michael Coates into the discussion as well.

Maisey replied that although he hadn't given the topic much consideration, he thought the issue might be one of insufficient preservation in the eugeneodontids more than an absence of claspers. Being the careful scholar he was, Maisey also noted the possibility that all the more complete eugeneodontid specimens could indeed be female. To be thorough, Maisey added that internal fertilization was possible in some cases without claspers, and that there was always the very remote possibility that eugeneodontids didn't practice internal fertilization. Having thus placed everything on the table, he circled back to what he viewed as the most likely option, that the fossils just weren't sufficiently preserved to reveal pelvic anatomy.

Coates agreed with Maisey. The eugeneodontids were simply too poorly known. Neither the case of the missing claspers nor the great gill debate would be settled until someone found a *Helicoprion* fossil with those parts preserved—and that's paleontology for you.

In a reflective moment at a different time, in a different discussion, around a campfire in Montana, Dick Lund mused that artists and paleontologists alike were left to reconstruct the enigmatic Paleozoic fish from scrap. Unfortunately, he said, they happened to be missing key pieces of scrap—or hadn't recognized the code-breaking bits in the thousands of pieces of scrap that legions of researchers had tossed aside over generations of digging and looking. Until someone found the missing pieces, any reconstructions beyond what was clearly evident from the scraps in hand were more in the realm of faith and speculation than science. And that was why it was important to keep digging. "Tomorrow!" he said. "We'll find the missing piece tomorrow!"

In the meantime, the show (and the showman) must go on. Art met science head-on in "The Whorl Tooth Sharks of Idaho" exhibit, which opened at the Idaho Museum of Natural History in June 2013. The walls were covered by Troll's *Helicoprion* art—in framed pieces, in a seventeen-foot-long mural on unframed canvas, and in floor-to-ceiling stylized images that Troll sketched out like fine-art wallpaper for painting-in by a squad of student volunteers from the fine arts, geosciences, and museum departments. A rolling video loop integrating footage produced by Ketchikan filmmaker Marc Osborne Jr. and enhanced by the IVL gang showed Troll with flaming whorls in his eyes, and featured an exploding tooth whorl—not to mention the animated segment where a *Helicoprion* bites off Troll's head.

Paleosculptor Gary Staab created two dramatic life-size replicas for the exhibit, one of which appeared to burst into the room through the wall, with chunks of sheetrock ripping away. (There was a tense moment when then-museum-director Herb Maschner, who had been out of town for the installation work, thought they had really punched a hole in the wall.)

Pruitt built an operating model of the jaw, complete with whorl. Instead of muscles, the model had gears to make the mouth open and close at the press of a button, and both kids and adults alike pressed, and pressed, and pressed. Theatrical gobo lights projected a spinning reproduction of Bendix-Almgreen's tooth whorl graphic the size of a kiddie pool on the floor, and a soundtrack played music from Troll's band, the Ratfish Wranglers, and also carried the voice of Rainer Zangerl, Troll's first fellow traveler on his spiraling *Helicoprion* path, recorded a few years before Zangerl's death.

Amid the eye-catching artwork, sculptures, chopping jaws, lights, and videos were fossils: Idaho no. 4 in a Plexiglas case resting on a bed of small white shark teeth, and in another case, the counterpart of 36701—the fossil that started Pruitt's "quick little research project," and one of the world's largest *Helicoprion* tooth whorls, measuring nearly two and a half feet in diameter with four and a quarter volutions. Also on display was 30897, the big chunk of whorl Pruitt showed Troll and the early-morning partiers in the parking lot at Caesars Palace in Las Vegas. The 30897 chunk was set next to a huge ammonite fossil that was even bigger around, while above them both hovered Staab's fifteen-foot-long, full-body *Helicoprion* replica, suspended from the ceiling.

The art made it fun, the fossils made it real. Each of those *Helicoprion* fossils—and all the others downstairs in the collections, and all the ones still lying hidden in rock layers around the world, had once been living, breathing animals, in a world full of other living, breathing animals. *What happened?* asked the art majors painting ghostly paleo chondrichthyans on the walls. *Where did they go?* asked the children pressing the button to make the model jaws open and close. It's a good question, with no good answer. No one knows for sure. The *Helicoprion* genus died out some twenty million years before the devastating mass extinction at the end of the Permian, leaving behind no clear explanation of their wholesale demise.

It's worth it here to draw a distinction between "background" extinction, like that of *Helicoprion*, and the accelerated extinctions of the so-called Anthropocene, the age of humans. Paleontologists estimate that most species naturally persist for one to ten million years, with the ten million mark being the exceptional upper end. It follows that, over the three and a half billion years of life on earth, lots of species have come and gone. Extinction is a biological fact of life, as is evolution. Species die out, and new species evolve. Researchers estimate that before humans appeared, and not counting mass extinction events, normal species loss was somewhere between .1 and 1 extinction per million species per year. Since the dawn of our own genus, *Homo*, those rates have risen *one hundred- to one thousandfold*, depending on how you count and who you ask. Endangered Species International reports that of the 44,838 species assessed worldwide by the respected International Union for Conservation of Nature (IUCN) Red List, 905 are extinct, with an additional 16,928 species listed as threatened to become extinct. From 1800 to 2015 alone, roughly the span of our whorl-tooth-fossil story, at least fifty vertebrate species have gone extinct, from the overhunted Steller's sea cow, to the ratted-out Tahiti sandpiper, to the poisoned-to-oblivion Hokkaido wolf, to Martha, the sole surviving passenger pigeon, to Lonesome George, the last known representative of the Pinta Island tortoise subspecies, who died in 2012. David M. Armstrong, a science teacher at the University of Colorado, Boulder, has described the acceleration of extinction rates as the difference between a casual drive and going Mach 2. It's alarming, distressing, heartbreaking, and, suggests Eileen Grogan, disrespectful to nature, which has given us so much.

By contrast, *Helicoprion* died, as it were, in its sleep. There were no such things as shark fin soup or organochlorines in the Permian, and the buzz saw sharks were allowed to play out a remarkably long run. The species life span of most higher vertebrates—animals like *Carcharodon carcharias* (the great white shark), or *Balaena mysticetus* (the bowhead whale), or *Homo sapiens*—typically averages about one million years. In comparison, the *Helicoprion* species plied their predatory trades for an impressive ten million years or more. The fossil record doesn't show any bigger, badder animal coming up to outcompete and displace them, and there isn't a clear marker that their prey base crashed. Neither does there appear to have been a calamitous planetary upheaval near the beginning of the mid Permian, around the time *Helicoprion* died out. Some scrappy eugeneodontids, whose fossils have been found in western Canada, survived the end-Permian extinction and eked their way into the Triassic, but the entire Eugeneodontida order was extinct by the close of the lower Triassic—taking with it *Helicoprion, Edestus, Sarcoprion, Caseodus, Fadenia, Ornithoprion*, and the rest. They left no direct heirs.

Extinction notwithstanding, *Helicoprion* was a bona fide success story. It was one of the largest predators of the time, at the top of its evolutionary game. Perhaps its success was part of its undoing. For one thing, *Helicoprion's* size might have begun to work against it. Full-grown adults were twenty to maybe thirty feet long. A carnivore that big has to eat a lot of meat to sustain itself. Even a minor dip in the food source could have set in motion an irreversible population decline. Which takes us to the next point in the potential problem column: their evolutionary masterpiece, the tooth whorl, quite likely painted them into the corner of extreme specialization. The tooth whorl must have been an ideal adaptation—as long as there was exactly the right kind and size of prey to use it on. Something must have changed however, to cause that amazing expression of genetic genius to lose its edge. When you need a screwdriver, pliers won't do. Such a highly innovated feature as a tooth whorl would be hard for evolutionary forces to remodel very quickly. There must not have been enough time for the eugeneodontids to shift gears and implement plan B.

Or maybe they fell prey to a bacterium, or virus, or parasite, or a change in ocean chemistry that was profound enough to kill off the *Helicoprion* genus, but too subtle to be captured in the fossil record. Yet. There's simply not enough information available, at least today, to know definitively what course of events took them down. Yes, there are holocephalans today, but only around fifty species, compared to the five hundred or so species of living sharks (plus more than six hundred additional species of skates and rays), and none that trace directly back to the eugeneodontids. The mind-boggling diversity of those early euchondrocephalans was bottlenecked at the Permian extinction, to be overtaken by their sister chondrichthyan subclass of "true" sharks.

Plenty of other *Helicoprion* life-and-death details remain unsettled as well, including who the buzz saw shark's most direct ancestor was, and how closely related, or not, *Helicoprion* and *Edestus* were.

———

The study of deep time requires a certain sangfroid. It demands an unflappable intellectual and even emotional disposition. Paleontology is not for the impatient. As Coates told Pruitt when the uninitiated undergrad was setting out on his *Helicoprion* quest, progress depends on sporadic discoveries of something new, something different, in fossils freshly unearthed from rock layers, or in already exhumed fossils dug from bottom shelves—or through the persistent probing of the perpetually curious.

In the spring of 2016, Pruitt was fine-tuning a new 3-D render he had been working on of a full-body *Helicoprion* computer model. When he put the jaws in—utilizing the technical, biological, and paleontological expertise he had gained over the preceding years to create as accurate a model as possible—he noticed that the throat opening ended up being tiny. *Hmm.* He reevaluated Idaho no. 4, modified the jaws a little bit, took out some lateral compression, and still ended up with a surprisingly small throat opening. In researching a number of different sharks, and found a correlation between the size of the throat and the size of the gill slits. *Hmmmm.* And with that, Pruitt took a step toward Lund and Grogan's opercular cover camp. He didn't grab his tent and move in, but he could see the campfire sparks over the fence.

Pruitt discussed what he'd found with Tapanila, and they decided to leave gill slits in the model for the time being. The two had resisted naming any new *Helicoprion* species out of the few anomalous specimens in their caliper-driven species study, and likewise, they wouldn't make a sudden move on gills, because the work and ideas were unfolding, waiting for more evidence, waiting for more pieces of the animal to show up. Those pieces had to be out there—fossil fragments that must be lying on the same bedding plane as their component tooth whorl, but had been customarily overlooked because the whorls were so dramatic. So obvious. So collectable. So mesmerizing.

Fifty miles southeast of the IVL, where Pruitt was finalizing his *Helicoprion* rendering, a colossal orange excavator beeped and clawed in the moonscaped maw of an open-pit phosphate mine. The mine's senior geologist, David Carpenter, went about his rounds as the rumbling excavator's giant steel bucket peeled back layers of time from the pit wall. Two hundred and seventy million years ago, Carpenter, the excavator—and Pruitt, Tapanila, and the waitresses at Elmer's— would have been at the bottom of the Phosphoria Sea. Carpenter, who happened to be working toward his master's degree in Tapanila's geoscience department at ISU, scanned the reject pile before heading back to the office. Something caught his eye. Whorls. Three *Helicoprion* whorls. Carpenter called Tapanila to let him know.

Yes! What else might be there? What other key bits and pieces of code-breaking scrap? *Keep looking. We'll find them.*

The *Helicoprion* torch was passing into good hands.

EPILOGUE

ONE OF THE MOST SURPRISING THINGS I LEARNED WHILE RESEARCHING AND WRITING this book was how unsettled the science of Paleozoic sharks is, and how few resources exist on the subject for general readers. The discovery of fossils that reveal new paleo shark information doesn't happen very often, but when it does, the resulting research can incite a flurry of rethinking. More consistent than finding new fossils is the rapid advance of revolutionary new tools and techniques for delving ever deeper into the details of ancient life, which begets a never-ending recalibration of facts and theories.

This is an exciting time for paleontology—and for paleo-buffs. You can go online right now, to sites like the Idaho Virtualization Laboratory and DigiMorph, and look at 3-D computer models of the skulls and skeletons of extinct animals. You don't need an advanced degree to ponder, investigate, and learn. Next time you drive by a road cut exposing millions of years worth of rock layers, or spot fossilized coral on a high ridge in the Rocky Mountains, or fork-up a shark tooth while digging potatoes in New Jersey, look for the questions lying just underneath. When? How? Why? What's the story? Our planet is a big, spinning ball of enormous amazement and intimate surprise. We can never solve all its mysteries or know the full scope of its endless, complicated workings, but we can wonder. We can look and dig and question.

Science fiction writer Isaac Asimov said something to the effect that, the most important expression in science isn't *Eureka*, but *That's odd* . . . Like a packed spiral of teeth in an animal that's supposed to shed its teeth. Some say the meek will inherit the earth, but really, it's the curious. It's yours for the asking.

ACKNOWLEDGMENTS

RESURRECTING THE SHARK RODE INTO BEING ON A GIANT WAVE OF GENEROSITY AND support. I owe a huge debt to Ray Troll for introducing me to *Helicoprion*, and for sharing his treasure trove of art, materials, memories, and connections spanning more than two decades of buzz saw obsession. This book exists first and foremost because of Ray. I am also deeply beholden to the rest of "Team Helico" and other members of the scientific community for helping me understand the often arcane and confusing finer points of paleontology, chondrichthyan phylogeny, and scanning technology. Leif Tapanila and Jesse Pruitt gave me countless hours of their time in interviews, and were always welcoming, accommodating, and patient with my endless questions. Jesse's offer to create augmented reality models for the book was beyond cool, and I can't thank him enough for his hard work and creative vision.

Cheryl Wilga cheerfully welcomed me into her then-home and saltwater lab in Rhode Island, where she showed me her bamboo sharks, educated me about jaw suspension, and shared her ideas while we sat on the rocks with our feet in Narragansett Bay. Thanks also to Dominique Didier and Alan Pradel for their time, stories, and expertise, and to Jason Ramsay for his explanation of jaw mechanics and his great shell-shucking illustration.

My abounding appreciation goes to Richard Lund and Eileen

Grogan for their warmth, forthrightness, and willingness to explain concepts and terms, then explain them again, and for their hospitality and lively conversation around the fire at their Montana field camp, with nighthawks flying overhead. I am also indebted to Oleg Lebedev, who through email correspondence from Russia graciously answered my questions about his *Helicoprion* work, and located and translated for me biographical information on Alexander Karpinsky, which was surprisingly scarce in the American literature.

I owe monumental thanks to my agent extraordinaire, Laurie Abkemeier, who encouraged me to stretch the original idea for the book, made me believe I could do it—then found the perfect home for the project in Pegasus Books. Pegasus associate publisher Jessica Case was incredibly supportive, collaborative, and a terrific pleasure to work with. I'm very grateful to Jessica for taking a chance on the book, and to designer Sabrina Plomitallo-González and the rest of the team for bringing it to such beautiful life.

For moral and literary support without measure I am eternally obliged to friend and fellow writer Susan Edsall, who read each chapter as I finished it. Her unbridled enthusiasm kept the snarling lions of self-doubt at bay. Penelope Pierce shored me up on difficult days and celebrated the incremental successes on walks with her dogs and my steadfast heeler, Teva. There was further lift-and-carry at every turn: more grounding dog walks with Maribeth Goodman, Suzanne Kingsbury's transcendent help with a preliminary book proposal, and the light on Kris Ellingsen's face when, early in the idea stage, I told her about this crazy beast and its quirky workers. "You have to write that book," she said.

Finally, abiding and heartfelt gratitude to my husband Rick—for eloping two months before the manuscript deadline, and for his love, support, patience, and unwavering sense of humor. Thank you, all.

ENDNOTES

p. 5   SCIENTISTS ESTIMATE THAT 99.9 PERCENT: "Foundational Concepts," Department of Paleobiology, National Museum of Natural History, Smithsonian Institution. http://paleobiology.si.edu/geotime/main/foundation_life4.html.

p. 5   ONE RECENT STUDY: Camilo Mora et al., "How Many Species Are There on Earth and in the Ocean?," in *PLOS Biology*, August 23, 2011. http://journals.plos.org/plosbiology/article?id=10.1371/journal.pbio.1001127.

p. 5   THE NUMBER OF EXTINCT SPECIES: Michael L. McKinney, "How Do Rare Species Avoid Extinction? A Paleontological View," in *The Biology of Rarity, Causes and Consequences of Rare-Common Differences*, Population and Community Biology Series 17, edited by W. E. Kunin and K. J. Gaston (New York: Springer-Science+Business Media, 1997), 110–111.

p. 7   ARTHUR RIVER: Henry Woodward writes in his *Geological Magazine* article that Mr. Davis found the fossil "in the valley of the Arthur River, an affluent of the Gascoyne from the right, i.e., the north, above the confluence of the Lyons with the Gascoyne." The most prominently known Arthur River in Western Australia is not in the Gascoyne watershed, but since Woodward specifically calls out the confluence of the Lyons, there must have been locally known Arthur River that was tributary to the Gascoyne.

p. 8   BEFORE THE CRYSTAL PALACE DINOSAUR COURT EXHIBIT OPENED: Joe Cain, "Dinner in the Iguanodon," Friends of the Crystal Palace Dinosaurs. http://cpdinosaurs.org/library/108.

p. 8   SCORN ANYONE ASSOCIATED WITH THE EXHIBIT: Steve McCarthy and Mick Gilbert, *The Crystal Palace Dinosaurs* (London: Crystal Palace Foundation, 1994), 85.

p. 9   IN 2015, SCIENTISTS RELEASED A THREE-HUNDRED-PAGE STUDY: Charles Choi, "The Brontosaurus is Back," *Scientific American*, April 7, 2015.

p. 10   TROLL AND MATSEN WERE COLLABORATING: The book was *Planet Ocean: A Story of Life, the Sea and Dancing to the Fossil Record,* by Brad Matsen with illustrations by Ray Troll (Berkeley, CA: Ten Speed Press, 1994).

2

p. 23   EVIDENCE: Martin Rudwick, *Earth's Deep History: How It Was Discovered and Why It Matters* (Chicago: University of Chicago Press, 2014), 163.

p. 23   AS A PROTESTANT BISHOP: Douglas O. Linder, "Bishop James Ussher Sets the Date for Creation," Famous Trials website. Ussher's date for Creation played a role in the famous Scopes trial, also known as the "Monkey Trial." At issue was the teaching of evolution in American classrooms. http://law2.umkc.edu/faculty/projects/ftrials/scopes/ussher.html.

p. 24   WHISPERS THAT THE WORLD WAS VASTLY OLDER: Claude C. Albritton Jr., *The Abyss of Time: Changing Conceptions of the Earth's Antiquity After the Sixteenth Century* (Mineola, NY: Dover Publications, Inc., 2002), 9.

p. 26   STENO DREW A SHARP DISTINCTION: Martin J. S. Rudwick, *The Meaning of Fossils: Episodes in the History of Paleontology,* 2nd ed. (Chicago: University of Chicago Press, 1976), 53.

p. 26   HE FURTHER ARGUED: ibid, 51.

p. 27   HOOKE INTRODUCED: Ben Waggoner, "Robert Hooke," University of California Museum of Paleontology, 2001. http://www.ucmp.berkeley.edu/history/hooke.html.

p. 28   "FAITHS AND CERTAINTIES OF CENTURIES PAST": Simon Winchester, *The Map That Changed the World: William Smith and the Birth of Modern Geology.* (New York: HarperCollins, 2001), 11.

p. 28   WHAT THE DAIRY MANAGERS CALLED: ibid, 29.

p. 29   OLD MINER'S TERM: Ian Duhig, in his poem "'Strata Smith," from *Pandorama* (London: Picador, 2011).

p. 29   HIS TEACHER AT THE VILLAGE SCHOOL: Winchester, *The Map That Changed the World*, 31.

p. 29   SMITH WALKED: ibid, 57.

p. 30   DIPPED SLIGHTLY DOWNWARD: ibid, 66.

p. 30   OBLITERATE THE POSSIBILITY: the one exception is volcanic tuff, a rock composed of volcanic ash. Fossiliferous tuff can be either igneous or sedimentary.

p. 31   310 AND 290 MILLION YEARS AGO: Winchester, *The Map That Changed the World*, 64.

p. 31   HE WOULD KNOW EXACTLY: ibid, 70.

p. 31   ITS OWN UNIQUE SET OF FOSSILS: ibid, 71–72.

p. 32   WINCHESTER CALLS MEARNS PIT AS SIGNIFICANT: ibid, 62.

p. 33   IN ADDITION TO A SCIENTIFIC ORGANIZATION: C. L. E. Lewis and S. J. Knell, *The Making of the Geological Society of London* (London: Geological Society Special Publication 317, 2009), 3.

p. 34   ERASMUS WAS A CARD-CARRYING "LUNATICK.": Winchester, *The Map That Changed the World*, 24.

p. 35   WRITING TO A FRIEND IN 1753: John Playfair, "Biographical Account of the late James Hutton, M.D.," in *The Works of John Playfair, Esq., Vol. IV*. (Edinburgh: Archibald Constable & Co., 1822), 40.

p. 35   The paper Hutton presented to the Royal Society of Edinburgh, called "Theory of the Earth," was published in Transactions of the Royal Society of Edinburgh in 1788. The paper was expanded and published in Edinburgh in 1795 as a two-volume book of the same name.

p. 35   HE READ ABOUT THE CHEMIST: Scottish Science Hall of Fame, "James Hutton." http://digital.nls.uk/scientists/biographies/james-hutton/.

p. 35   HE SIMPLY CONSIDERED IT LARGELY IRRELEVANT: George Gaylord Simpson, "Uniformitarianism. An inquiry into Principle, Theory, and Method in Geohistory and Biohistory," in *Essays in Evolution and Genetics in Honor of Theodosius Dobzhansky*, edited by Max K. Hecht and W. C. Steere (Amsterdam: North Holland, 1970), 47.

p. 36   TOASTED MICE TO ROASTED RHINO: David Allan Feller, "Zoophagous geology: William Buckland and extra-visual scientific observation," University of Cambridge lecture, November 2010. http://talks.cam.ac.uk/talk/index/26827.

p. 36   WAS FAMOUSLY SAID: "Who wants to eat jellyfish omelets? Dolphin meatballs? Mouse-on-toast? These guys," Krulwich Wonders, Robert Krulwich on Science, National Public Radio, September 27, 2012. http://www.npr.org/sections /krulwich/2012/09/27/161874316/who-wants-to-eat-jellyfish-omelettes-dolphin -meatballs-mouse-on-toast-these-guys.

p. 37   KINDRED SPIRIT: Sandra Herbert, *Charles Darwin, Geologist* (New York: Cornell University Press, 2005), 184.

p. 37   WAS VEHEMENTLY OPPOSED: John van Whyhe, "Georges Cuvier, leader of elite French Science," The Victorian Web. http://www.victorianweb.org/science/cuvier.html.

p. 37   DELICATELY BALANCED: Edward J. Larson, *Evolution: The Remarkable History of a Scientific Theory* (New York: Modern Library Chronicles, reprint edition 2006), 21.

p. 38   THE EMBODIMENT OF REASON: ibid., 10.

p. 40   FROM HER LETTERS: Betty Gilderdale, *The Seven Lives of Lady Barker* (Christchurch, NZ: Canterbury University Press, 1996), 271–272.

p. 41   ONE EARLY VISITOR WROTE: ibid., 275.

p. 41   IN WHICH SHE DESCRIBES: Lady Barker, *Station Life in New Zealand* (London: Macmillan and Co., 1871), 175.

3

p. 46   JUDICIOUS AND TACTFUL EDITOR: Obituary, Henry Woodward. *The Geological Magazine* 58 (November 1921): 481–484.

p. 46   TWO OF HIS DAUGHTERS: Susan Turner, et al., "Forgotten Women in an Extinct Saurian (Man's) World," *Geological Society, London Special Publications* (October 2010) 135–138.

p. 46   IN 1868, HENRY WOODWARD WROTE: Obituary, Henry Woodward.

p. 47   WHILE TRAVELING TOGETHER IN SWITZERLAND: "Sir Philip Grey Egerton, 10th Baronet," *Wikipedia.* https://en.wikipedia.org/wiki/Sir_Philip_Grey _Egerton,_10th_Baronet.

p. 47   LAUNCHED THE FIELD: Hans-Peter Schultze, editor's preface to *Handbook of Paleoichthyology, Volume 3D, Chondrichthyes,* by Michal Ginter, et al. (Munich: Verlag Friedrich Pfeil, 2010), 5.

P. 48   INSTALLED ELECTRIC LAMPS: "The Electric Light in the British Museum," *New York Times,* December 18, 1879. http://query.nytimes.com/mem/archive-free/pdf? _r=1&res=9A0CE1DC163EE63BBC4052DFB4678382669FDE&oref=slogin.

p. 48   "I READILY IDENTIFIED": Henry Woodward, "On a Remarkable Ichthyodorulite from the Carboniferous Series, Gascoyne, Western Australia." *The Geological Magazine,* Vol. III, No. I (January 1886): 1.

p. 48 "I HAVE BEEN REQUESTED": ibid., 7.

p. 54 ROAM THE BANKS: "Joseph Leidy (1823–1891)," Penn University Archives & Records Center, Penn Biographies. http://www.archives.upenn.edu/people/1800s /leidy_joseph.html.

p. 54 JOSEPH WAS SO INCORRIGIBLY TRUANT: Leonard Warren, *Joseph Leidy, the Last Man Who Knew Everything* (New Haven, CT: Yale University Press, 1998), 23.

p. 54 TURNIPS: W. S.W. Ruschenberger, MD, *A Sketch of the Life of Joseph Leidy, M.D., LL.D.,* (Philadelphia: MacCalla, 1892).

p. 57 TOLL TAKEN BY THE CIVIL WAR: Warren, *Joseph Leidy*, 132.

p. 57 INCLUDING THE ONE ON *H. FOULKI*: ibid., 139.

p. 57 SCULPTOR BENJAMIN WATERHOUSE HAWKINS: Robert McCracken Peck, "The Art of Bones," adapted from *All in the Bones: A Biography of Benjamin Waterhouse Hawkins*, by Valerie Bramwell and Robert M. Peck, published by the Academy of Natural Sciences, 2008, and reprinted in *Natural History* with permission from the academy. Excerpt published online at http://www.naturalhistorymag.com /features/11340/the-art-of-bones?page=2.

p. 58 FOR HIS BEHAVIOR: Warren, *Joseph Leidy*, 189.

p. 58 IN A PAPER PUBLISHED: ibid., 176.

p. 59 AS SOON AS HE READ: ibid., 177.

p. 59 "IN PALEONTOLOGY": ibid., 192.

p. 59 WITH FEW OR NO GUIDING LANDMARKS: ibid., 202.

p. 59 FORT GIBSON HAD BEEN ESTABLISHED: US Department of the Interior, National Park Service, Fort Gibson National Cemetery, Fort Gibson, Oklahoma; nps.gov website. http://www.nps.gov/nr/travel/national_cemeteries/Oklahoma /Fort_Gibson_National_Cemetery.html.

p. 60 AN ESTABLISHMENT NEAR FROZEN ROCK: Carl C. Branson, "Type Species of *Edestus* Leidy," *Oklahoma Geology Notes*, a newsletter of the Oklahoma Geological Survey, University of Oklahoma 23, No. 12 (December 1963): 275–276.

4

p. 62 THE FIRST LINE OF THE ENTRY READ: Joseph Leidy, "Indications of Five Species, with

Two New Genera, of Extinct Fishes," *Proceedings of the Academy of Natural Sciences of Philadelphia, v. 7* (October 1855): 414.

p. 62   "NO VERTEBRATED ANIMAL": Joseph Leidy, "Descriptions of Some Remains of Fishes from the Carboniferous and Devonian Formations of the United States," *Journal of the Academy of Natural Sciences of Philadelphia*, ser. 2, v. 3 (1855–1858): 159.

p. 62   "IT IS A FAIR INFERENCE": ibid.

p. 63   IN HIS DRAWING: Wayne M. Itano, "A Tale of Two Holotypes: Rediscovery of the Type Specimen of *Edestus minor,*" *Geological Curator* 10 (2014): 19.

p. 64   "A BAD FIGURE OF IT": John Strong Newberry, *Paleozoic Fishes of America,* a monograph of the United States Geological Survey, Volume XVI, Government Printing Office (1889): 218. Browse the whole book, a personal gift of Louis Agassiz to Harvard, at https://archive.org/stream/paleozoicfisheso00newb#page /n1/mode/2up.

p. 64   "IT MAY PERHAPS BE": Leidy, "Description of Some Remains of Fishes from the Carboniferous and Devonian Formations of the United States," 302.

p. 64   "FOSSIL FRAGMENT OF THE JAW": ibid., 159.

p. 65   FOR WHICH HE ERECTED: Itano, "A Tale of Two Holotypes," 19.

p. 66   ILLINOIS STATE GEOLOGIST: Charles A. White, "Memoir of Amos Henry Worthen, 1813–1888" (read before the National Academy of Sicences, November 1893), 345. http://www.nasonline.org/publications/biographical-memoirs/memoir-pdfs/ worthen-amos.pdf.

p. 66   THE YOUNGEST OF NINE CHILDREN: *A History of the First Half-Century of the National Academy of Sciences 1863–1913* (Baltimore: The Lord Baltimore Press, 1913), 164–165. https://books.google.com/ books?id=SddRAAAAMAAJ&printsec=frontcover&source=gbs_ge_ summary_r&cad=0#v=onepage&q&f=false 164.

p. 66   THERE WERE FISH: Charles A. White, "Biographical Memoir of John Strong Newberry, 1822–1892" (read before the National Academy of Sciences, April 17, 1902), 5. http://www.nasonline.org/publications/biographical-memoirs /memoir-pdfs/newberry-j-s.pdf 5.

p. 67   AS FAR AS ITS GREAT CAÑONS: ibid.

p. 67   FROM THE DIVISION'S HEADQUARTERS: ibid.

p. 67   IN A POSTWAR ACCOUNTING: Michael A. Flannery, *Civil War Pharmacy: A History of Drugs, Drug Supply and Provision, and Therapeutics for the Union and Confederacy* (New York, London, Oxford: Pharmaceutical Products Press, 2004), 62.

p. 67   TWENTY-SIX-YEAR CAREER: Frederick W. True, ed., *A History of the First Half-Century of the National Academy of Sciences 1863–1913* (Baltimore: Lord Baltimore Press, 1913), 166.

p. 68   THE EARLIEST IDENTIFIABLE: Ginter, *Handbook of Paleoichthyology*, 19.

p. 68   LEIDY'S FOSSIL: John Strong Newberry and A. H. Worthen, "Descriptions of Fossil Vertebrates," *Geological Survey of Illinois, Volume IV, Geology and Paleontology, Section I* (published by the authority of the Legislature of Illinois, 1870), 350.

p. 69   IN HIS EPIC: John Strong Newberry, *Paleozoic Fishes of America* (a monograph of the United States Geological Survey, Volume XVI, Government Printing Office, 1889), 220. Browse the whole book, a personal gift of Louis Agassiz to Harvard, at https://archive.org/stream/paleozoicfisheso00newb#page/n1/mode/2up.

p. 70   LIKELY SHE ENROLLED: Richard Lund, personal communication, July 14, 2015.

p. 70   WOMEN'S ATHLETIC PROGRAM: Mark Frazier Lloyd, "Women at Penn: Timeline of Pioneers and Achievements 1880–1900." http://www.archives.upenn.edu/histy/features/women/chron3.html.

p. 70   THE EDUCATION OF WOMEN: "Fanny Rysam Mulford Hitchcock (1851–1936)," Penn University Archives & Records Center, Penn Biographies. http://www.archives.upenn.edu/people/1800s/hitchcock_fanny.html.

p. 71   "TO PURSUE A HIGHER EDUCATION": ibid.

p. 71   AS WAS ANNA LEIDY: "Portraits of Joseph Leidy," The Academy of Natural Sciences, Ewell Sale Stewart Library. http://www.ansp.org/research/library/archives/99900-99999/leidy9/.

p. 71   SHE FOUND SOME IMPORTANT FOSSILS: Warren, *Joseph Leidy*, 146.

p. 72   QUITE PLAUSIBLE: John Strong Newberry, *Paleozoic Fishes of America* (a monograph of the United States Geological Survey, Volume XVI, Government Printing Office, 1889), 222. Browse the whole book, a personal gift of Louis Agassiz to Harvard, at https://archive.org/stream/paleozoicfisheso00newb#page/n1/mode/2up.

p. 73   "CERTAINLY SUCH A MONSTER": ibid., 219–222.

p. 73   "ATTACK AND DEFENSE": ibid., 224.

5

p. 76   ONE OF ABOUT SIXTEEN: Ben Eklof, *Russian Peasant Schools: Officialdom, Village Culture, and Popular Pedagogy, 1861–1914* (Los Angeles and Berkeley, University of California Press, 1990), 134–135.

p. 76   A SECRET MEETING IN MINSK: "First Congress of the Russian Social Democratic Labour Party," *Wikipedia*, https://en.wikipedia.org/wiki/1st_Congress_of_the _Russian_Social_Democratic_Labour_Party.

p. 77   LENIN'S OWN FATHER: Eklof, 129.

p. 77   HAD A TEACHER SELECT: ibid., 137.

p. 78   IMPERIAL RUSSIAN GEOLOGICAL SURVEY: A. S. Woodward, "Reviews—Karpinsky on *Helicoprion*" *(Geological Magazine,* Decade IV, Vol. VIII, (January–December: 1900), 34.

p. 78   COPPER MINING SETTLEMENT: "Krasnoturyinsk," https://en.wikipedia.org/wiki /Krasnoturyinsk.

p. 79   UNFLINCHING COURAGE: "Ural-batyr," *Wikipedia*, https://en.wikipedia.org/wiki /Ural-batyr.

p. 79   IN THE END, URAL-BATYR CONQUERED DEATH: "Folklore of the Bashkirs" http:// kulturarb.ru/en/literature/.

p. 79   FABERGÉ: "Ural Mountains," *Wikipedia*, https://en.wikipedia.org/wiki/Ural _Mountains.

p. 79   IMPERIAL LAPIDARY FACTORY: Will Lowes and Christel L. McCanless, Fabergé Eggs: A Retrospective Encyclopedia (Scarecrow Press, 2001), 207.

p. 80   MINISTRY OF TRUTH: S. I. Romanovsky, *Aleksandr Petrovich Karpinsky, 1847– 1936* (Leningrad: Nauka Publishers, 1981); taken from notes from personal communication with Oleg Lebedev, May 26, 2014.

p. 80   HIS RESEARCH AND IDEAS: Irina V. Batyushkova, "Karpinsky, Alexandr Petrovich" *Complete Dictionary of Scientific Biography*, 2008. http://www.encyclopedia.com/ doc/1G2-2830902258.html.

p. 82   LIKE OWLS SNATCHING MICE: Neale Monks and Philip Palmer, *Ammonites* (Washington: Smithsonian Institution Press in association with the Natural History Museum, London, 2002), 55.

p. 84 *DOLIODUS PROBLEMATICUS*: Sean Markey, "World's Oldest Shark Fossil Found," *National Geographic News*, October 1, 2003. http://news.nationalgeographic.com/news/2003/10/1001_031001_sharkfossil.html.

p. 84 REVOLVER DENTITION: Ginter et al., *Handbook of Paleoichthyology*, 16.

p. 88 CONVINCING OURSELVES: Roderick Impey Murchison et al., *The Geology of Russia in Europe and the Ural Mountains, Vol. I*, (London: John Murray, Albemarle Street, 1845), 138, 141.

p. 89 URALIC LANGUAGE FAMILY: "Permic Languages," Wikipedia, https://en.wikipedia.org/wiki/Permic_languages.

p. 89 *PERAMA*, MEANING "FARAWAY LAND": RussiaTrek.org, http://russiatrek.org/perm-city.

p. 89 PERM WAS SO REMOTE: "Perm," Russia-InfoCentre. http://www.russia-ic.com/regions/3229/3230/history/.

p. 90 A DIVERSITY OF SPECIES: The Marine Realm and the End-Permian Extinction, Smithsonian Department of Paleobiology. http://paleobiology.si.edu/geotime/main/htmlversion/permian4.html.

p. 91 TWENTIETH-CENTURY PALEONTOLOGIST CURT TEICHERT: Curt Teichert, quoted by D. H. Erwin in "The End-Permian Mass Extinction," in *The Permian of Northern Pangea: Volume 1: Paleogeography, Paleoclimates, Stratigraphy*, edited by Peter A. Scholle et al. (Berlin and Heidelberg, Germany: Springer-Verlag, 1995), 20.

p. 93 THE ANTHROPOCENE: Author Elizabeth Kolbert outlines the evidence and arguments surrounding the Anthropocene in her compelling book *The Sixth Extinction: An Unnatural History* (New York: Henry Holt, 2014).

6

p. 97 HE BELIEVED THE SPIRAL GROWTH: Irina V. Batyushkova, "Karpinsky, Alexandr Petrovich," http://www.encyclopedia.com/doc/1G2-2830902258.html.

p. 99 IN A THIN SECTION: Karpinsky ordered his thin sections from the German company Voigt und Hochgesang, according to S. I. Romanovsky, *Aleksandr Petrovich Karpinsky*, as translated and shared in personal communication with Oleg Lebedev, May 26, 2014.

p. 100 KARPINSKY FELT THE EVIDENCE: Irina V. Batyushkova, "Karpinsky, Alexandr Petrovich," http://www.encyclopedia.com/doc/1G2-2830902258.html.

p. 100   DID HIS BEST: S. I. Romanovsky, *Aleksandr Petrovich Karpinsky,* as translated and shared in personal communication with Oleg Lebedev, May 26, 2014.

p. 102   ONE OF THE MOST COLORFUL PAGES: S. I. Romanovsky, *Aleksandr Petrovich Karpinsky,* as translated and shared in personal communication with Lebedev, May 26, 2014.

p. 102   ERNEST VAN DEN BROECK: Van den Broeck is usually cited in the literature as "Van Den Berg."

p. 102   A GOOD GEOLOGIST: Biographical information about van den Broeck taken from Geert Vanpaemel, "The Global and the Local: The history of science and the cultural integration of Europe." *Proceedings of the 2nd International Conference of the European Society for the History of Science (ICESHS),* Cracow, Poland: September 6–9, 2006.

p. 103   ALSO APPEARING: As reported in Eastman's literature review, C. R. Eastman, "Literature of *Edestus,*" *The American Naturalist* 39 (June 1905): 405–409. http://www.jstor.org/stable/2454909?seq=5#page_scan_tab_contents.

p. 103   FEW WILL BE PREPARED: C. R. Eastman, "Karpinsky's Genus *Helicoprion,*" *The American Naturalist* (Chicago: University of Chicago Press for American Society of Naturalists, July 1900), 579–582.

p. 104   HIGHLY SPECIALIZED OFFSHOOTS: Eugene Willis Gudger, ed., *The Bashford Dean Memorial Volume, Archaic Fishes.* (New York: published by order of the Trustees, American Museum of Natural History, 1930), 13. https://archive.org/stream/bashforddeanmemo01dean#page/6/mode/2up.

p. 104   AS PUZZLING AS EVER: Eastman, "Karpinsky's Genus *Helicoprion,*" 579–582.

p. 104   A MORE CONCORDANT RESPONSE: A. S. Woodward, "*Helicoprion*—Spine or Tooth?" *Geological Magazine* (Decade IV) 7, No. 1 (January 1900): 33–36.

p. 105   JUST AFTER HIS EIGHTEENTH BIRTHDAY: Susan Turner and John Long, "The Woodward Factor: Arthur Smith Woodward's legacy to geology in Australia and Antarctica," *Geological Society Special Publication,* Lyell Collection, published online October 28, 2015.

p. 105   THE MONUMENTAL THREE-VOLUME *CATALOGUE*: A. S. Woodward, *Catalog of the fossil fishes in the British Museum.* (London: Printed by order of the Trustees, 1889–1901). http://catalog.hathitrust.org/Record/001489474.

p. 105   SCOTTISH PALEONTOLOGIST R. H. TRAQUAIR: Traquair's Devonian fossil was *Protodus scoticus,* which has since been identified as *Nostolepis,* an extinct genus of acanthodian as described by Carole J. Burrow and Susan Turner: "Reassessment of 'Protodus' scoticus from the Early Devonian of Scotland," *Morphology, Phylogeny and Paleobiogeography of Fossil Fishes,* David K. Elliott, John Maisey, et al., eds (Munich: Verlag Friedrich Pfeil, 2010), 123–144.

p. 107 EVOLUTIONARY SCIENCE: "The History of Popular Science," *Popular Science* online, July 23, 2002. http://www.popsci.com/scitech/article/2002-07/history-popular-science.

7

p. 111 "BUT VERITABLE TEETH": C. R. Eastman, "Some Carboniferous Cestraciont and Acanthodian Sharks," *Bulletin of the Museum of Comparative Zoology at Harvard College* 39, No. 3 (June 1902): 56.

p. 112 INTERPRETATION WAS AT FAULT: Ibid., 65.

p. 112 AS THESE TEETH: Ibid., 65.

p. 113 AN INDEX OF ALL THE PAPERS: C. R. Eastman, "Literature of *Edestus,*" *The American Naturalist* 39 (June 1905): 405–409. http://www.jstor.org/ stable/2454909?seq=1#page_scan_tab_contents.

p. 114 F. JOHN.: This *Helicoprion* illustration is often attributed to the famous landscape and paleoartist of the same time period, Heinrich Harder, who also illustrated Wilhelm Bölsche articles.

p. 114 NAMED ANOTHER NEW: Oliver Perry Hay, "A new genus and species of fossil shark related to *Edestus* Leidy," *Science* 26 (July–December 1907): 22–24.

p. 116 WOULD LIKELY HAVE REEKED: Leif Tapanila, personal correspondence, October 29, 2015.

p. 117 MOST PHYTOPLANKTON: "What are Phytoplankton?," National Oceanic and Atmospheric Administration, National Ocean Service website. http://oceanservice. noaa.gov/facts/phyto.html.

p. 117 ONE HUNDRED TO ONE: Leif Tapanila, personal correspondence, October 29, 2015.

p. 118 PHOSPHATES ARE PRIMARILY MINED: Phosagro Annual Report 2011. http://ar2011 .phosagro.ru/eng/ingredients/phosphorus/.

p. 118 HAY DESCRIBED: Oliver Perry Hay, "On the Nature of *Edestus* and Related Genera, with Descriptions of One New Genus and Three New Species," *Proceedings of the U.S. National Museum* 37 (1910): 43–61.

P. I2I   FIVE ADDITIONAL FOSSIL FRAGMENTS: "Notes," *Nature* (September 21, 1916): 54. https://books.google.com/books?id=-1ZGAQAAMAAJ&p g=PA54&lpg=PA54&dq=Helicoprion+clerci&source=bl&ots=h3beP ZqHSD&sig=j9JYNtTUioxwlmmeVBGD8K3Ykqo&hl=en&sa=X &ved=0ahUKEwi7oN--qNfKAhUW9GMKHdaMDiUQ6AEIM jAF#v=onepage&q=Helicoprion%20clerci&f=false.

p. I22   SHOULD BE POLITICALLY AND FINANCIALLY SUPPORTED: "Alexander Karpinsky," *Wikipedia.* http://en.wikipedia.org/wiki/Alexander_Karpinsky.

p. I22   COMMISSIONED AS A: Donald J. LaRocca, "Bashford Dean and Helmet Design During World War I," Metropolitan Museum of Art, online, About the Museum (July 23, 2014). http://www.metmuseum.org/about-the-museum/now-at-the-met/2014/bashford-dean-and-helmet-design-during-world-war-i.

8

p. I28   ONE OF JUST SIX: Rex E. Crick and George D. Stanley Jr., "Curt Teichert (1905–1996)," AAPG Bulletin (March 1997), Memorials, American Association of Petroleum Geologists. http://archives.datapages.com/data/bull_memorials/81/081003/pdfs/494.htm.

p. I28   THEY REMAINED: "Biographical Material on Curt Teichert," Oral History Project, K U Retiree's Club, University of Kansas, 1991. http://www.kuonlinedirectory.org/endacott/data/OralHistoryTranscripts/TeichertCurt.pdf.

p. I28   MR. H. COLEY PICKED UP: Curt Teichert, "Helicoprion in the Permian of Western Australia," Journal of Paleontology 14 (1940): 141.

p. I28   WHEN ANOTHER ROCKEFELLER GRANT: ibid.

p. I29   ENABLED TEICHERT TO TAKE: John A. Reinemund, "Memorial to Curt Teichert, 1905–1996," Geological Society of America, Memorials 28 (Nov. 1997): 40. ftp://rock.geosociety.org/pub/Memorials/v28/teichert.pdf.

p. I29   THE GOVERNMENT SEEMED SURPRISED: "Teichert, Curt," *Complete Dictionary of Scientific Biography*, 2008. Encyclopedia.com. (June 28, 2016). http://www.encyclopedia.com/doc/1G2-2830906138.html.

p. I30   TEICHERT WAS ABLE TO CONTINUE: The couple emigrated to the United States in 1952, where Teichert became an esteemed teacher at the University of Kansas, and later served as an adjunct professor of geological sciences at the University of Rochester until 1993. He died in 1996 at the age of ninety-one. "Biographical Material on Curt Teichert," Oral History Project, K U Retiree's

Club, University of Kansas, 1991. http://www.kuonlinedirectory.org/endacott/data/OralHistoryTranscripts/TeichertCurt.pdf.

p. 130   EVEN TODAY, NO ANSWER: Curt Teichert, "Helicoprion in the Permian of Western Australia," Journal of Paleontology 14 (1940): 144.

p. 130   SMALL TOOTH-LIKE ELEMENTS: Walter Youngquist, personal communication, August 20 and 22, 2014.

p. 130   WITH COLONIES ESTABLISHED IN 1721: William James Mills, Exploring Polar Frontiers: A-L, Vol. 1 (Santa Barbara, CA: ABC-CLIO, 2003), 273. https://books.google.com/books?id=PYdBH4dOOM4C&pg=PA358&lpg=PA358&dq=eigil+niel sen+denmark+paleontologist&source=bl&ots=hdE2BKxyll&sig=tuKZfFA6EQ7y_EBo5UQofmiR1bA&hl=en&sa=X&ved=0ahUKEwjAwsLR2IHLAhUBxGMKH QNAC2cQ6AEIHzAB#v=snippet&q=greenland&f=false.

p. 137   MOST OF A SHARK'S SKELETON: Mason N. Dean et al., "Ontogeny of the Tessellated Skeleton," *Journal of Anatomy* 215 (2009): 227–239. http://www.ncbi.nlm.nih.gov/pmc/articles/PMC2750757/.

p. 137   WITH EXTRA LAYERING: Anthony P. Farrell et al, ed., *Encyclopedia of Fish Physiology: From Genome to Environment* (Academic Press, Cambridge, MA: 2011).

p. 137   WHICH MEANS, LONG SAYS: John Long, "No bones about it, sharks evolved cartilage for a reason," published online in *The Conversation*, May 28, 2015. http://theconversation.com/no-bones-about-it-sharks-evolved-cartilage-for-a-reason-42258.

9

p. 146   A WARTHOG-LIKE ANIMAL: Brian Switek, "Charles R. Knight's Prehistoric Visions," Smithsonian.com, Jan. 6, 2012. http://www.smithsonianmag.com/science-nature/charles-r-knights-prehistoric-visions-16099537/?no-ist.

p. 146   FROM THAT BEGINNING: Marianne Sommer, "Seriality in the Making: The Osborn-Knight Restorations of Evolutionary History," *History of Science* 48 (2010): 461–482.

p. 147   WRITE THE BOOK: Rainer Zangerl, *Handbook of Paleoichthyology, Volume 3A, Chondrichthyes I, Paleozoic Elasmobranchii* (Stuttgart, Germany, and New York: Gustav Fischer Verlag, 1981).

p. 149   FOR WHICH BODY FOSSILS DID EXIST: John Maisey, Curator-in-Charge, Division of Paleontology at the American Museum of Natural History, points out that science knows the body form in only one of Zangerl's eugeneodontids, which wasn't in the group that included *Helicoprion*. From Maisey personal communication, July 27, 2016.

p. 149   LEADING SCHOLARS: John Maisey, personal communication, July 27, 2016. "We could lump Port Jackson sharks with stingrays on the basis that they both have dentitions consisting of large crushing teeth. But we would be completely wrong (classically, the same wrong comparison was made with Port Jackson sharks and extinct hybodonts from the Mesozoic). We know the body form in only one of Zangerl's kinds of 'eugeneodontids,' and it isn't the group that includes *Helicoprion*. So if we didn't know what a Port Jackson shark looked like, we would get a very odd idea by looking at a stingray. Zangerl's 'kit and caboodle' is extremely dubious."

p. 152   HIS ARGUMENTS: Zangerl, *Handbook of Paleoichthyology*, 82.

10

p. 161   LEBEDEV STUMBLED: Oleg Lebedev, personal communication, July 14, 2015.

p. 162   "CURIOUSLY": W. W. Nassichuk, "Permian Ammonoids in the Arctic Regions of the World," in *The Permian of Northern Pangea*, edited by Peter A. Scholle et al. (Berlin and Heidelberg, Germany: Springer-Verlag, 1995), 210.

p. 162   OBRUCHEV HAD RECENTLY PUBLISHED: D. V. Obruchev, "Edestid studies and the works by A. P. Karpinsky," *Trudy Paleontologicheskogo Instituta Akademii Nauk SSSR*, 45:1–85. (In Russian.)

p. 163   AN ADDITIONAL PIECE: Richard Glenn, "An Alaskan *Helicoprion* from Atigun Gorge, East-Central Brooks Range, Alaska," presented at the Geological Society of America Cordilleran Section 111th annual meeting, May 2015.

p. 166   AUSTRALIAN PALEONTOLOGIST JOHN LONG: John A. Long, *The Rise of Fishes* (Baltimore and London: Johns Hopkins University Press, 1995), 76.

p. 166   A HIGHLY EFFECTIVE ARRANGEMENT: Oleg Lebedev, "A new specimen of *Helicoprion* Karpinsky, 1899 from Kazakhstanian Cisurals and a new reconstruction of its tooth whorl position and function," *Acta Zoologica* 90 (May 2009): 178.

p. 170   LONG AND MAISEY HAD VENTURED: John Long, personal communication, June 18, 2015.

11

p. 180   FRIENDS EVER SINCE: Dominique Didier, personal communication, September 21, 2015.

p. 181   "EVERY INVESTIGATOR WILL ADMIT": Bashford Dean, *Chimaeroid Fishes and Their Development* (Washington: Carnegie Institution of Washington, 1906), 5, 151, 155.

p. 181   HE HAD AN "ARDOR": William Gregory, "Memorial of Bashford Dean, 1867–1928," in *The Bashford Dean Memorial Volume, Archaic Fishes,* published by order of the Trustees of the American Museum of Natural History (1930–1933): 5.

p. 189   ANGELA MILNER CARRIED: David Perlman, "Study Shows Dinosaur Could Fly," *SFGate,* published online August 5, 2004. http://www.sfgate.com/news/article/Study -shows-dinosaur-could-fly-Winged-creature-2703140.php.

p. 190   PAVED OVER BY A WALMART: Alan Pradel, personal communication, August 21, 2015.

12

p. 197   ARTICLE IN *NATURAL HISTORY*: Adam Summers, "When the Shark Bites," *Natural History,* published online, March 2006. http://www.naturalhistorymag.com /biomechanics/172103/when-the-shark-bites.

p. 197   SOMEONE WAS PLAYING A JOKE ON HER: Cheryl Wilga, personal communication, July 27, 2015.

p. 200   EARLY JAWLESS FISH: Susan Turner and Randall Miller, "New Ideas About Old Sharks," *American Scientist,* May–June 2005.

p. 200   MIRED IN CONTROVERSY: P. C. Donoghue and M. Rücklin, "The Ins and Outs of the Evolutionary Origin of Teeth," *Evolution & Development,* 18 (2016): 19–30; epub September 15, 2014. http://www.ncbi.nlm.nih.gov/pubmed/25219878.

p. 201   RECENT PAPERS HAVE ARGUED: ibid.

p. 202   AT THE TIME, GROGAN WAS A "WHITE LAB COAT" BIOLOGIST: Eileen Grogan, personal communication, July 14, 2015.

p. 205   EPIPHANIES THAT HE SHARED WITH PRUITT: Robert Schlader, personal communication, January 10, 2014.

p. 211   JOHN MAISEY PUBLISHED A PAPER: John G. Maisey, "What is an 'elasmobranch'? The impact of paleontology in understanding elasmobranch phylogeny and evolution" *Journal of Fish Biology* 80 (2012): 918–951.

13

p. 220   THE FIRST PAPER: Leif Tapanila et al., "Jaws for a Spiral-tooth Whorl CT Images

Reveal Novel Adaptation and Phylogeny in Fossil *Helicoprion*," *Biology Letters* 9 (February 2013).

p. 220   TAPANILA AND PRUITT'S SPECIES PAPER: Leif Tapanila and Jesse Pruitt, "Unraveling Species Concepts for the *Helicoprion* Tooth Whorl," *Journal of Paleontology* 86 (2013): 965–983.

p. 221   THE THIRD TEAM PAPER: Jason Ramsay et al., "Eating with a saw for a jaw: functional morphology of the jaws and tooth-whorl in *Helicoprion davisii*," *Journal of Morphology* (January 2015): 47–64; (epub September 2014).

p. 222   JOHN LONG TELLS THE STORY: John Long, *The Dinosaur Dealers* (New South Wales, Australia: Allen & Unwin, 2002): 53–57.

p. 223   LEBEDEV SAID THE PAPER: Oleg Lebedev, personal communication, May 26, 2014.

p. 224   "FROM THIS VERSION": Augustus (Toby) White, "Holostyly: the Suspension Mounts," Chondrichthyes: Holocephali, palaeos.com http://palaeos.com /vertebrates/chondrichthyes/holocephali.html.

p. 225   THE FOSSIL EVEN PRESERVED SKIN PIGMENTS: Eileen Grogan and Richard Lund, personal communication, July 14, 2015.

p. 225   *DEBEERIUS* TO SUPPORT THEIR ARGUMENT: Eileen Grogan and Richard Lund, "Debeerius Ellefseni (Fam. Nov., Gen. Nov., Spec. Nov.), an Autodiastylic Chondrichthyan from the Mississippian Bear Gulch Limestone of Montana (USA), the Relationships of the Chondrichthyes, and Comments on Gnathostome Evolution," *Journal of Morphology* (2000): 219–45. http://www .ncbi.nlm.nih.gov/pubmed/10681469.

p. 225   GROGAN *ET AL*: Eileen Grogan, Richard Lund, and Dominique Didier, "Description of the chimaerid jaw and its phylogenetic origins." *Journal of Morphology* 239 (1999): 45–59. Holocephali.

p. 226   AGAIN DRAWING: Augustus (Toby) White, "Holostyly: the Suspension Mounts," Chondrichthyes: Holocephali, palaeos.com http://palaeos.com /vertebrates/chondrichthyes/holocephali.html#Holocephali.

p. 229   IN THE BIOLOGY LETTERS PAPER: Leif Tapanila et al., "Jaws for a Spiral-tooth Whorl.

p. 229   AUTHOR BRIAN SWITEK: Brian Switek, "Buzzsaw Jaw *Helicoprion* Was a Freaky Ratfish," Laelaps blog in Phenomena: A Science Salon, National Geographic online, Feb. 2013. http://phenomena.nationalgeographic.com/2013/02/26 /buzzsaw-jaw-helicoprion-was-a-freaky-ratfish/.

p. 234   GLIDING JOINTS: Robert E. Shadwick et al., eds, *Physiology of Elasmobranch Fishes: Structure and Interaction with Environment*, (Academic Press, an imprint of Elsevier, 2016), 161. https://books.google.com/books?id=oS1OBQAAQBA J&pg=PA161&lpg=PA161&dq=gliding+joint+in+sharks&source=bl&ots=2-g HTKABnE&sig=OqxO-g2mnptFg5oEIy5shpHJMsA&hl=en&sa=X&ved=0a hUKEwjAia7T24fNAhVRyGMKHbS1ACcQ6AEIPjAF#v=onepage&q=glid ing%20joint%20in%20sharks&f=false.

p. 238   ALSO HOW FAST: This can be explained very simply, says Ramsay, by Newton's laws of motion. Newton's second law basically says force equals mass times acceleration (F = ma). If we rearrange that equation to a = F/m, it shows that if mass (m) doesn't change, then acceleration (a) increases with increasing forces (F). More force = more acceleration of the jaw = a faster bite! Jason Ramsay, personal communication, August 25, 2015.

p. 239   "CHAIN ARMOR": Oleg Lebedev, "A new specimen of *Helicoprion* Karpinsky, 1899 from Kazakhstanian Cisurals and a new reconstruction of its tooth whorl position and function," *Acta Zoologica* 90 (May 2009): 178.

p. 241   IT SWAM: John Maisey, personal communication, July 27, 2016.

|4

p. 245   AS TOBY WHITE WROTE ON PALAEOS.COM: Augustus (Toby) White, "Holostyly: the Suspension Mounts," Chondrichthyes: Holocephali, palaeos.com http://palaeos.com/ vertebrates/chondrichthyes/holocephali.html.

p. 246   ALL MATURE MALE: Jeffrey Carrier, John A. Musick, and Michael R. Heithaus, eds. *Biology of Sharks and Their Relatives*, 2nd ed. (Boca Raton, FL: CRC Press, 2012).

p. 246   IN 2009: Per Ahlberg et al., "Pelvic claspers confirm chondrichthyan-like internal fertilization in arthrodires." *Nature* 460 (2009): 888–889. http://www.nature.com/ nature/journal/v460/n7257/abs/nature08176.html.

p. 247   LONG HAD WRITTEN: John Long, *The Dawn of the Deed* (Chicago: University of Chicago Press, 2012).

p. 247   HE CIRCLED BACK: Expanding on the clasper subject, Maisey notes that chimaeras have claspers, as do elasmobranchs. "Phylogenetic bracketing," he explains, "suggests that 'euchondrocephalans' had them too. There is still a big question about when claspers actually evolved; they are absent in acanthodian fishes, which are today considered to be on the stem lineage leading to chondrichthyans. It isn't known if the earliest 'conventionally defined chondrichthyans' (i.e., those with the special 'tessellated calcified cartilage') had claspers. But it seems that by the time the elasmobranch and holocephalan lineages diverged, claspers had appeared." John Maisey, personal communication, July 27, 2016.

p. 248   DICK LUND MUSED: Dick Lund, personal communication, July 14, 2015.

p. 248   "THE WHORL TOOTH SHARKS OF IDAHO EXHIBIT": Once the exhibit started traveling, the name was changed to "Buzz Saw Sharks of Long Ago."

p. 250   BETWEEN .1 AND 1: J. M. De Vos et al., "Estimating the normal background rate of species extinction," *Conservation Biology* 29, No. 2 (April 2015): 452. http://www.ncbi.nlm.nih.gov/pubmed/25159086.

p. 250   ONE HUNDRED- TO ONE THOUSANDFOLD: "Extinction: A Natural versus Human-Caused Process," Russell Labs, University of Wisconsin–Madison, (2011). http://labs.russell.wisc.edu/peery/files/2011/12/7.-Extinction-a-Natural-and-Human-caused-Process.pdf.

p. 250   ENDANGERED SPECIES INTERNATIONAL REPORTS: Endangered Species International organization website. http://www.endangeredspeciesinternational.org/overview1.html.

p. 250   DAVID M. ARMSTRONG. A SCIENCE TEACHER: David M. Armstrong, "Point-Counterpoint: Why Worry about Extinction," in *Human Biology*, 7th ed., Daniel D. Chiras (Burlington, MA: Jones & Bartlett Learning, 2012), 484.

p. 250   DISRESPECTFUL TO NATURE: Eileen Grogan, personal communication, July 15, 2015.

INDEX